BODY ENGINEERING

The Institution of Mechanical Engineers

Proceedings 1969–70 · Volume 184 · Part 3M

BODY ENGINEERING

A Symposium arranged by the
Automobile Division
of the Institution of Mechanical Engineers
and the Advanced School of Automobile Engineering

1 BIRDCAGE WALK · WESTMINSTER · LONDON S.W.1

ISBN 0 85298 046 9

CONTENTS

		PAGE
Introduction		vii
Opening Lecture		1
Paper 1	The realism of vehicle impact testing, by C. R. Ennos, C.Eng., M.I.Mech.E.	6
Paper 2	The effects on body structures of present and possible future safety legislation, and the mathematical simulation of barrier impact, by J. Curtis	19
Paper 3	A study of impact behaviour through the use of geometrically similar models, by G. W. Barley, B.Sc.(Eng.) (*Graduate*), and B. Mills, Ph.D., C.Eng., M.I.Mech.E.	26
Paper 4	Towards the all-plastics motor car, by G. O. Gurney	34
Paper 5	Windscreens of the future, by S. E. Kay	44
Paper 6	Body components, by A. E. Norman	50
Paper 7	Structural analysis of car body shells using computer techniques, by G. G. Moore, B.Eng., Ph.D.	55
Paper 8	Small computer procedures as tools for structural designers, by G. A. Wardill	62
Paper 9	The structural design of bus bodies, by G. H. Tidbury, B.Sc., C.Eng., F.I.Mech.E.	71
Paper 10	Automotive design with special consideration for safety in interior design, by J. E. Fallis	84
Paper 11	Chassis frames, by D. W. Sherman	88
Paper 12	Human factors influencing control positions, by G. R. W. Simmonds	96
Paper 13	Automobile body testing techniques, by W. R. Greenaway	102
Paper 14	The experimental investigation of body structural vibration, by M. Rodger, B.Sc. (*Graduate*)	108
Paper 15	Sound damping control of automobile body structures, by A. M. Chappuis	126
Paper 16	Modern methods of translating a styling model to a body draft and tooling, by D. W. Davy (*Graduate*)	134

PAGE

Discussion 142

Authors' replies 150

List of delegates 155

Index to authors and participants 157

Subject index 158

Body Engineering

A SYMPOSIUM was held at the College of Technology Cranfield, Bedfordshire, from the 6–7th July 1970. It was sponsored by the Automobile Division of the Institution of Mechanical Engineers and the Advanced School of Automobile Engineering, 159 delegates registered to attend. The Symposium was formally opened by D. A.Chilver, Ph.D., M.Sc., M.A., C.Eng., Vice Chancellor of the Cranfield Institute of Technology. There followed an Opening Lecture by C. M. Mackichan.

The papers were divided into seven sessions for presentation and discussion.

Monday, 6th July

SESSION 1: Styling of the 70s.
Chairman: H. R. Barber, C.Eng., M.I.Mech.E., Opening Lecture.

SESSION 2: The effects of international safety regulations on future structural body design.
Chairman: J. H. Alden, C.B.E., C.Eng., F.I.Mech.E., Papers 1, 2 and 3.

SESSION 3: Body components.
Chairman: A. Piaziali, B.S.(Mech.Eng.), Papers 4, 5 and 6.

SESSION 4: Stress and deflection prediction techniques by digital computers.
Chairman: H. Sheron, C.Eng., M.I.Mech.E., Papers 7, 8 and 9.

Tuesday, 7th July

SESSION 5: Interior styling and layout.
Chairman: P. Wilkes, Papers 10 and 12.

SESSION 6: Body testing techniques.
Chairman: R. H. Macmillan, M.A., C.Eng., F.I.Mech.E., Papers 13, 14 and 15.

SESSION 7: General problems
Chairman: C. S. King, Papers 16 and 11.

The members of the Organizing Committee were: J. H. Alden (Chairman), H. R. Barber, C.Eng., M.I.Mech.E., J. R. Ellis, Ph.D., M.Sc.(Eng)., C. R. Ennos, C.Eng., B. B. Hundy, B.Sc., Ph.D., K. A. Osborne, G. M. Palmer, G. H. Tidbury, B.Sc., C.Eng., F.I.Mech.E., G. A. Wardill and J. A. B. Wolfe.

Opening Lecture

THE CHALLENGE OF AUTOMOBILE DESIGN FOR THE 1970s

C. M. MacKichan*

Although we have long been concerned with functional considerations in automobile design, there is an ever-increasing, sometimes conflicting, series of demands upon the automobile designer today. Some of the problems outside aesthetics which face the designer in the 1970s are outlined and the ways in which GM styling plans to adopt advanced technology to make the designer's job more efficient. The final section deals with the designer's role in society. The designer is a true artist who reflects, and even anticipates, the conditions of the society which surrounds him.

THE INDUSTRY is being challenged from all sides in terms of ecological and politico-social concerns. If exhaust emissions are not the subject, then it is passenger safety, traffic congestion or consumer satisfaction.

But theoretical consideration of these twentieth century problems is not enough. One can't be a theorist about safety, for example. We must test theories on the proving grounds. The air-bag restraint system you hear so much about today may be obsolete tomorrow . . . for we learn more about our business each year.

Let it be said we're not ducking any of these problems as they relate to our task. The designer's job must be to relate these challenges to the design of a good-looking car. It must be saleable; safe—yes, but not ugly, for good visual design and safety are certainly compatible.

As designers, we know the automobile, its relation to the driver and how it fits into the transport system. We were applying the principles of functional design years before the government began to take an active interest in automobiles. The fact is, however, that legislation is very much with us.

In America, when the ground rules of a situation change completely, we have an expression which says, 'It's a whole new ballgame'. And let me assure you, automobile design for the 1970s is a *whole new ballgame*.

We regard this situation as a tremendous challenge. We see the task of designing automobiles to meet the many detailed functional requirements as an opportunity to exercise our considerable skills and ingenuity to an even greater degree during the coming decade. We want to go on record as saying that automobiles in the 1970s will offer the customer a blend of beauty and function which far surpasses anything we have seen in the past.

I would like to describe for you some of the requirements that have been placed upon the functional aspects of the automobile in recent times, and some that we foresee being applied in the future. And then I would like to tell you about some of the sophisticated tools we are developing to enable us to do the job more rapidly and efficiently.

Contrary to complaints expressed by some of our critics, automobile designers have always been concerned with the functional aspects. We have to get our passengers in and out of cars, have to seat them comfortably and they have to be able to see out *over* the bonnet. We have been concerned with ramp and departure angles at the front and rear of the car, adequate lighting and many, many other functional considerations.

It was at our urging that cars were lowered, the passengers were located between the frame rails instead of above them, and that running boards were eliminated. The result was a lower, wider automobile with a lower centre of gravity, better handling and, in all, a better product. Though our motivations were frequently aesthetic, a more functional product often resulted. It's been said that what *looks* better is *indeed* better.

Since the coming of the Motor Vehicle Safety Standards in 1968 and our corporation's determination to equal or better these standards, the functional considerations of the automobile have been dissected, reconsidered, and

The MS. of this paper was accepted for publication on 3rd July 1970.

* *General Motors Building, Staff-Warren, Detroit 2, Michigan, U.S.A.*

restructured in a rather rigid form. Often, the requirements placed upon a particular component of the automobile conflict with desirable characteristics for another part of the automobile. It wouldn't serve the purpose of this discussion to detail *every* area in which the functional requirements of the automobile have been rigidly prescribed but it may be enlightening to concern ourselves with some of them at this time.

On the exterior of the automobile, one of the things requiring a major portion of our attention in the 1970s is bumper protection. As our society has become more congested, its concern no longer lies with such frivolities as three-tone paint schemes and chromium decoration but rather in areas of consideration for the pocketbook. The current emphasis on bumper protection is a reflection of this. In addition, it affects the safety aspects for, with increased bumper function, we provide protection for the lamps, the bonnet latches, the trunk latches and fuel fillers. In other words, we help preserve some of the safety in minor accidents.

All of this is well and good but there are some conflicts. The position of the lighting elements on the car is now rather strictly prescribed. The requirements for cooling the engine may also conflict with increased bumper protection.

This means the designer must exercise his creativity within narrower limits. Park/signal lights must be outboard; headlights must be 24 in off the ground and visible from a prescribed cut-off angle; bumper protection must be extended over a vertical range and he'd better not forget to leave enough holes in the front of the car for cooling the engine. Here again, theory is insufficient and all of these functional considerations should be studied on proving grounds before being accepted as gospel.

The problem is no less complicated at the rear of the car. Increased bumper protection there involves moving the lights and licence plates out of the bumper and doing something else with the filler pipe, for the hole that it now occupies is a void in the armour at the back of the car. The relocation of the filler pipe is of considerable importance because we have to consider fuel security; a front impact in an accident must not uncouple the tank. Rear and side impacts must not squash the filler pipe or disconnect it, with resultant petrol spillage and fire.

Increased congestion inflicts its toll on the side of the automobile. When we decreased the amount of chromium trim on the car in an effort to make it appear cleaner and more elegant, we quite frequently omitted the side strip with the result that we began to get increasing customer complaints. The anonymity of the car user in American cities, the increased shopping-centre parking, with its side-by-side accommodation for automobiles, have resulted in a lot of minor damage to the sides. The dings and dents suffered are unsightly and expensive to repair. So now we are faced with the problem of providing side protection as well, and all of our cars currently offer this as an option. It is a challenge to our skill and ingenuity to incorporate more side protection into the cars and still retain the flexibility to provide product differentiation and attractive appearance for our various automobiles.

As if this weren't enough, we have the problem of over-the-bonnet vision. After the requirements for a band of protection, a band of cooling, and a band of lighting have been met, we still have to be able to see over the top at some downward angle.

Rearward vision is a function of both interior and exterior design and it, too, will play a great part in the future. The concern for it is strongly reflected in the models we are bringing out in the early 1970s. The concern with vision and protection is a challenge to us to develop new approaches to the exterior of the car.

Functional requirements on the interior of the car are no less rigidly prescribed, both by government and by our own corporation. In fact, because the driver and passengers of the car are more intimately involved with the interior, it could be said that there are even more constraints for interior design than there are for exterior design. Our designers are subject to the criteria of serviceability, assembly (that is, within the car), the ease with which pieces come together, suitability of materials, and safety, in addition to competitiveness in terms of appearance.

Today, we are very concerned with the ability for service people to have access to items in the interior of the car which need regular maintenace. We are designing instrument panels which are removable so that components can be replaced; light bulbs can be exchanged as they burn out. We have designed heater units that can be exchanged for service and have provided access panels to other areas of functioning equipment in the instrument panel.

The instrument panel on a modern American automobile is no simple fascia board that can be slipped into place with ease. It is a large, bulky affair which has to be threaded into the body as it is being trimmed. How to get this bulk in and around other components requires a study of the sequence of operations. Thus, the assembly procedure must be integrated into the design process.

In terms of buildability, the ease of how the various components come together, how one material will act with another, whether or not the design can be built for a reasonable cost, are other areas of concern, subject to some constraints.

All of these factors are very closely interrelated with the aspects of safety. Safety presents a broad range of constraints in the interior of the automobile and, should it be found that some aspect of the design directed toward serviceability, assembly or buildability is in conflict with a safety requirement, then the process must be completely re-examined in order that the design meet the safety requirement.

All areas that are in danger of being hit in collisions have special conditions applied to them. For example, if the instrument panel can take a force of $80g$ in 3 ms but the panel knobs cannot, then the knobs themselves are subject to a series of restrictions—they must have rounded edges, be recessed or, if not, they must break away; or, if they can't be broken off, they must project no more than 3/8 in beyond the surface.

When a design has been established, it is tested with a device we have developed, called 'Trauma-saf'. This is a mechanical approximation of the human head which measures the amount of injury that would be expected under impact conditions.

In addition to the physical characteristics of the interior components, there are other areas of constraints. Vision requirements are established by our corporation. They have to do with up and down and left and right forward vision, affecting the design of the instrument panel, the windshield pillar and the bonnet. To the rear, we are also concerned with upward vision, downward vision, and to the left and right to eliminate blind spots.

To ensure that good forward vision is maintained, the government has divided the windshield into three areas and prescribed the percentage of wiping action which must be provided for each of these areas. The most critical area is the portion directly in front of the driver. We have developed some elaborate methods to ensure that reflections will not occur in that area.

Automotive designers are even involved in the sound level of their interiors. Much of the work is directed toward developing a quiet environment for the driver and selection of material and shapes plays a part here. This is in the interest of promoting greater safety through the reduction of irritation and subsequent fatigue; also by making it easier for the driver to hear outside warning devices, such as police and fire sirens, at times when he has the windows rolled up and his air conditioner on.

The legibility of instruments and controls is of extreme importance to us. We take into consideration that many of our drivers wear bi-focal glasses. Thus the size and frequency of gradations are of concern. We also study instrument size, colour and intensity of lighting and provide them with anti-reflective lenses. In addition, there are requirements for the location and standardization of controls, established both by the corporation and by government. All these controls must be within the reach of what we call a Class 2 belt system, which includes a seat belt and shoulder harness.

Farther down the road there are requirements under study for a passive restraint system which will protect the driver and passengers in the event of a collision. I refer, of course, to the 'air bag', a subject which is, at the moment, fraught with unanswered questions. We feel that, in the event this becomes a reality, it will have as great an influence on the interior packaging of the car as anything in the past.

With all the constraints imposed, the design must look good and it must be competitive. It must meet or exceed the efforts of other groups of designers working under the same constraints at Ford, Chrysler, and General Motors. All our designers are facing this challenge and we are proud of our ability to design well under all of the applied constraints.

We are subject to much more corporate review at the present time than we have been in the past. We have first, the Forward Product Planning Group. If it approves our ideas, they are then subject to review by our Engineering Policy Group. Other relevant committees within the corporation are the General Technical Committee and the Safety Review Board and various subcommittees. So the designer is now doing his job with a lot more people looking over his shoulder.

Now that I have outlined the many external constraints within which the automobile designer has to work and which will become a way of life for him in the 1970s, I would like to project some of the advanced tools that he will have at his command.

Throughout the 1960s we have developed a number of devices to make the designer's task easier and to shorten the time required to bring his idea to three-dimensional reality.

We have developed some rather sophisticated electronic measuring devices. Rather than take templates laboriously from a full-size clay model, we have developed an electronic surface recorder which scans the model and instantly converts the information to section drawings. We have also developed a point-taker which translates the surface of the automobile into digital information which is printed out and punched on numerical control tapes, which then can be fed into an automatic drafting machine. This dramatically shortens the time required to go from model to drawing. But as regards tools for the designer himself advances have been relatively slight.

We have adopted such modern conveniences as aerosol spray cans of paint, and the designer has learned to draw full-size on our blackboard drawings by using black tape, which can be lifted, re-arranged and re-contoured very rapidly. Nevertheless, up to now, by and large, he has been working just about as he had done for the past two decades.

This, we believe, will change during the 1970s. Some very useful devices are under development which use the computer's outstanding speed and ability to handle huge quantities of information; and capitalize upon its ability to create pictures on a television screen with electronic impulses. We have, in its infancy, a new system which will provide a very sophisticated tool for the designer. This consists of what we call a 'dynamic sketch pad' and a 'light pen' with which the designer can draw. With this new tool, he is in complete control of the process and can call up information concerning his car design on a television-like console before him. He can then transpose it, move it, or originate information which goes into the computer to add to the body of knowledge about this particular car design.

This very complex but extremely useful system will enable him to take into consideration all the many engineering inputs and all of the restrictive constraints placed upon the design. Thus he cannot become confused or forget important criteria when he begins to lay out a new automobile. He can inspect the design at close range or examine what goes on underneath it by drawing on the huge 'data bank' of information which has been put into the computer by the car divisions. This concerns chassis, frame, suspension, and the like. For example, he can simulate the dynamic action of a door-swing to determine exit and entry, panel interference and kerb clearance and can vary the hinge points to achieve a satisfactory design.

Today, when we initiate a new design, we lay out the car through a long, tedious process of pulling out drawings from the file, making tracings of localized areas, superimposing them and finally putting them together to make a composite drawing of a vehicle. We take suspension from one layout, engine and drive train from another; eventually we put in the seats, and the 'Oscar', or human manikin, comes into play. We try to put all the components into a compromise arrangement and every time we have to change something by trial and error, the task becomes time-consuming and tedious. Simulation compresses the time required.

Using his graphic console and light pen, coupled to the computer, the designer can very quickly arrange things to a more satisfactory configuration and he can look at many more possibilities in a given period of time. So we think that dynamic simulation is going to become the designer's primary tool in the creation of an automobile. It eliminates the need to build models or prototypes at this early stage and gives a greater degree of assurance that the design will effectively meet all requirements.

We realize that the designer couldn't possibly do effective appearance design on a TV screen. What he *can* do is get the basic layout of the automobile organized in his graphic console and then request that a full-size drawing be made to serve as an underlay for his full-size blackboard drawing. Now the traditional process begins and he works over the blackboard drawing until he arrives at what he feels is a desirable appearance. Then the blackboard drawing can be returned to the drawing machine, the function of which can be reversed to become a scanner, or line-follower. The information which the designer has generated then goes back into the computer.

With the updated mathematical model in the computer, the designer can now obtain a perspective representation of the lines he has drawn on the blackboard. He can see how the design fits the various constraints and he should be in a better position to determine whether or not it should be further developed as a full-size clay model.

This system, which is called CADANCE, an acronym for Computer Aided Design and Numerical Control Effort, is under development and has been demonstrated to be quite workable. We are in the process of accruing a huge data bank of information to make this possible and at the same time we are arranging the system so that the designer himself need bring to it no more knowledge or skills than he has already developed. In other words, the machine will be as easy to use as dialling a telephone is today.

The primary use of such a system as CADANCE is to reduce the time between statement of a design and its execution. The graphic console and the computer will team up to give the designer a much easier access to the amount of information that he needs to digest or examine and possibly reject. All the safety parameters, all of the protection constraints could be shown to him at the time he initiates his design. These new criteria will become as fundamental as our manikins have been in the past.

We expect the CADANCE programme to be functioning on a practical basis early in 1973.

However, we do not foresee that the computer will in any way replace the designer. There can be no effective substitute for creativity and judgement. The use of this extremely valuable system will put a very sophisticated tool in the hands of the designer, one which will extend his capability and enable him to work faster and more accurately, without retracing his steps.

So far we have been talking about a direct aid to the designer in the development of his creativity. Now we should consider some of the equally sophisticated tools which will reduce the time required to bring a design to fruition, or enable the design management to evaluate more proposals than is now possible.

Our clay modellers are trained artists, most of whom hold fine arts degrees. At the present time they must work at some rather dull, repetitive tasks. Once a design of a full-scale model has been completed on one side, it must be reproduced on the opposite side so that it can be properly evaluated. This tying up of the time and energy of truly creative people can be reduced by machine reproduction of a full-size model from NC data either generated in the computer or taken from the finished side of the model. We believe it is quite possible and will be feasible within the coming decade to maintain a master model at some place other than the creative studio, which will represent accurately both sides of a clay model made by creative people. This model itself will probably not be made of clay and it will be almost entirely generated by a machine. It will, of course, require hand finishing to bring it to the state of perfection necessary for proper evaluation. This will enable design managers to bring corporate management in for review of a project without interrupting normal activities in a creative studio. There is no target date for this type of system, but the technology exists and it appears to be reasonably feasible.

Beyond this—thinking even farther into the future—there is the possibility that full-size, three-dimensional images of cars may be generated electronically. While this sounds like science fiction, the technology exists today on a smaller scale. So it seems entirely possible that some time in the late 1970s design managers may be able to look at a holographic image in three dimensions before deciding if it should become a solid model.

These very advanced tools, which will enable the designer to go about his task rapidly and efficiently, may seem to be 'way out' but please recall that at the beginning of the 1960s we had not put a man in space. The things which I have just described are nowhere near so complicated as the task of putting a man on the moon. Sophisticated, yes, but tools that we will come to regard as indispensable before this decade is past.

Our response to the challenges and opportunities of the 1970s can result in dramatic changes to the automobile. However, we are still in an evolutionary business and we must first develop designs, assess the results and then move progressively to greater achievements.

It is apparent that the role of the designer is expanding, not contracting, as he adapts to the increasing concerns of society. The complexity of his task grows as the interdisciplinary nature of all human efforts affects his way of life. The designer is no less of an artist, but the exercise of his art becomes more demanding. He *alone* is at ease with the emotional aspects of automobile design and *alone* can ensure the aesthetic nuances that constitute attractiveness in a product, establish its visual character and express its function.

The designer faces a new and exciting future in our ever-changing society. Just as artists down through the centuries have done, the automobile designer will rise to the challenges posed by society and technology and will make an ever-increasing contribution to the quality of the civilization he serves.

Paper 1

THE REALISM OF VEHICLE IMPACT TESTING

C. R. Ennos*

Clearly it is easier and more convenient to test a vehicle's crash characteristics against a static or mobile barrier than against another car. However, it does not always yield the same results and a number of tests, discussed here, had the aim of correlating results from the various types of test. On the whole front and rear impacts can be simulated by static barriers but side impacts are much more difficult.

INTRODUCTION

THE ADVENT of motor vehicle safety legislation over the past five years has led to the organized destruction of literally thousands of motor vehicles. These have been sacrificed in an attempt to determine their impact performance with the ultimate goals of:

(i) Meeting safety legislation requirements;
(ii) Improving the occupant protection afforded by the vehicle.

Numerous test and analysis methods have now been evolved to simulate the various types of real-life accident which occur. The major part of this work has taken place in the United States, where many papers have been published on crash dynamics. Very little work, however, has been presented on the impact characteristics of the smaller European vehicle. On both sides of the Atlantic published data on actual test results are still rare.

A comprehensive vehicle impact test programme is currently being carried out by the Ford Motor Company in Europe to evaluate test methods and generally assess the impact performance of the small European vehicle.

This paper presents and discusses some of the initial results of this research programme and, where possible, comments on the validity of the test methods employed.

TEST AND INSTRUMENTATION TECHNIQUES

The majority of the standard vehicle impact tests are relatively simple to perform, once some means of propelling the test vehicles has been found. The method employed by Ford in Europe uses a servo-controlled winch motor, driving a continuous loop of cable. This permits impacts of vehicles with barriers or other vehicles, both moving and stationary, as shown in Fig. 1.1.

Whilst the tests are relatively simple, if expensive, to conduct, the associated instrumentation is complex and can become the limiting factor of the test. A minimum instrumentation requirement for a vehicle-to-vehicle impact would comprise five high-speed cameras ($\simeq$1000 f/s), 18 channels of accelerometers, six channels of seat belt tensometers, and six channels of event markers.

The instrumentation problem is twofold. The initial problem of ensuring that the equipment can accurately record the 100–200 ms crash transients becomes one of how to maintain this accuracy during the analysis. Where filtering of the transducer signals is employed during the analysis, extreme care must be exercised. How often does one see vehicle impact performance compared on a peak-acceleration basis? All too often the duration of the acceleration is not quoted, and neither are the response characteristics of the measuring equipment. The peak '*g*' levels quoted in this way are meaningless; by varying the frequency-response characteristics of the recording equipment it is possible, for a given impact, to obtain almost any '*g*'-value between zero and infinity.

Frequency response of transducer signals

To illustrate the significance of frequency response, Fig. 1.2 shows a typical vehicle deceleration trace successively filtered: note the changing '*g*' levels. The question may well be asked, why filter accelerometer traces at all? The main reason is that the requirements to ensure realistic comparisons between accelerometer recordings are not certain. The tendency therefore, is to record with a relatively high frequency response to avoid missing something that may be significant. After all, unwanted signals can always be filtered.

The MS. of this paper was received at the Institution on 17th April 1970 and accepted for publication on 1st May 1970.
* *Manager, Engineering Research—Body, Ford Motor Company Limited, Laindon, Basildon.*

Fig. 1.1. A servo-controlled continuous loop of cable produces a controlled vehicle-to-barrier impact in this Ford test facility

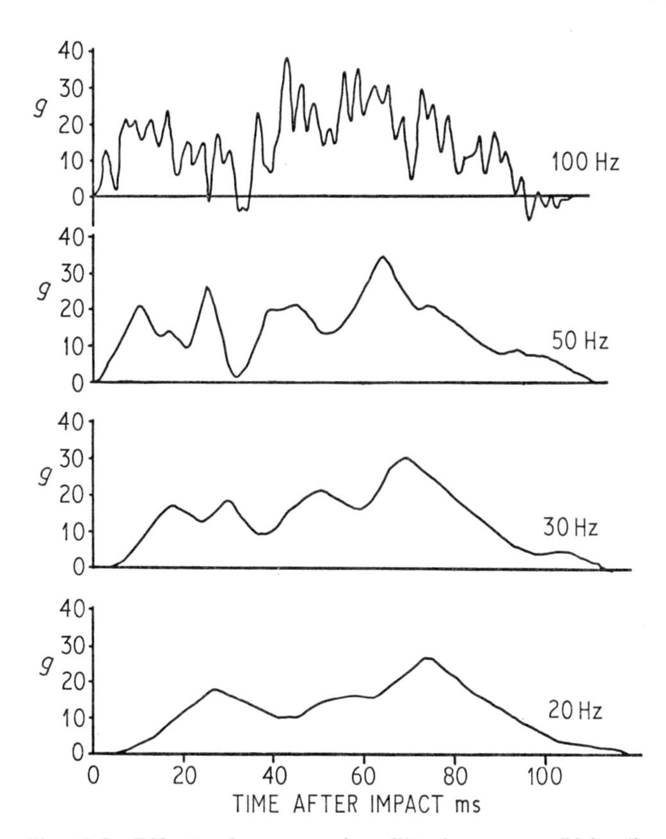

Fig. 1.2. Effect of progressive filtering on a 50 km/h frontal barrier impact on a typical deceleration curve

Basically, the aim is to compare the deceleration levels at various points on the vehicle and its dummy occupants, with one or other and with similar measurements on other vehicles. Unfortunately, vibration over a wide spectrum of frequencies is generated by the impact. The signal sent by the accelerometer derives from local structural ringing and electrical noise as well as the actual deceleration pulse of the vehicle.

This pulse, for a 50 km/h barrier impact, has a period of about 100 ms, corresponding to a frequency of 10 Hz. The transient, however, is not a smooth sine pulse; it is often double peaked, hence a 20 Hz equivalent frequency is more realistic. It is possible, however, that frequencies above this figure are important when comparing impact performance between vehicles. With regard to dummy occupant accelerometers it is necessary to resolve the traces to at least 1·0 ms to enable the Severity Indices (relating deceleration to occupant damage) to be calculated.

To ensure adequate data retrieval, a dual system is employed by Ford where the crash transients are recorded on both magnetic tape and UV galvanometer recorder systems possessing flat responses to both 2000 Hz and 100 Hz. A typical UV trace for a body-mounted accelerometer is given in Fig. 1.3, from which it is evident that there is still considerable vibration content at frequencies well in excess of 100 Hz. This is mainly due to structural ringing between 300–500 Hz.

Hand-filtering of these UV traces is, therefore, still necessary before they can be used to compare overall

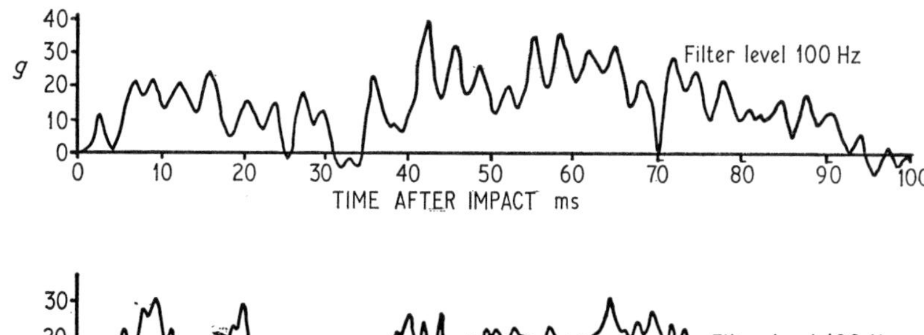

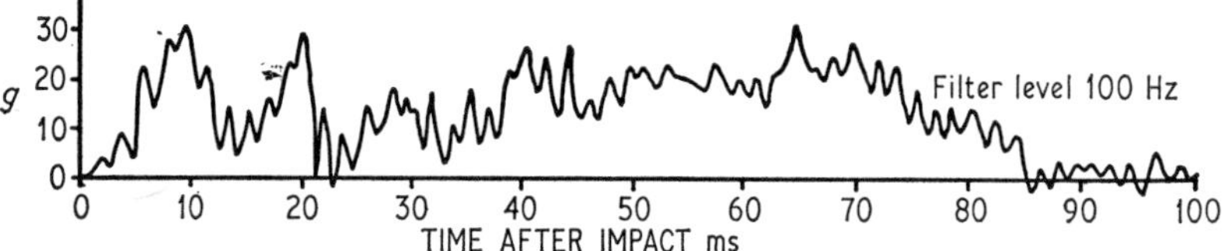

Fig. 1.3. Typical traces from body-mounted accelerometers for 50 km/h frontal impacts: there is considerable vibration even at more than 100 Hz

vehicle decelerations. The magnetic tape recording can be filtered to any desired level. This, however, also has its attendant problems which are related to the type of filter employed.

Fig. 1.4 illustrates the time-delay characteristic through a typical active filter for symmetric and asymmetric double-peak triangular waveforms on time bases similar to the typical crash transient. It is immediately evident from this figure that, at filter frequencies below 80–100 Hz, considerable real-time slippage occurs which affects both the start time and the slope of the pulse. This means that considerable care must be exercised in relating filtered acceleration signals to other measurements on a time basis.

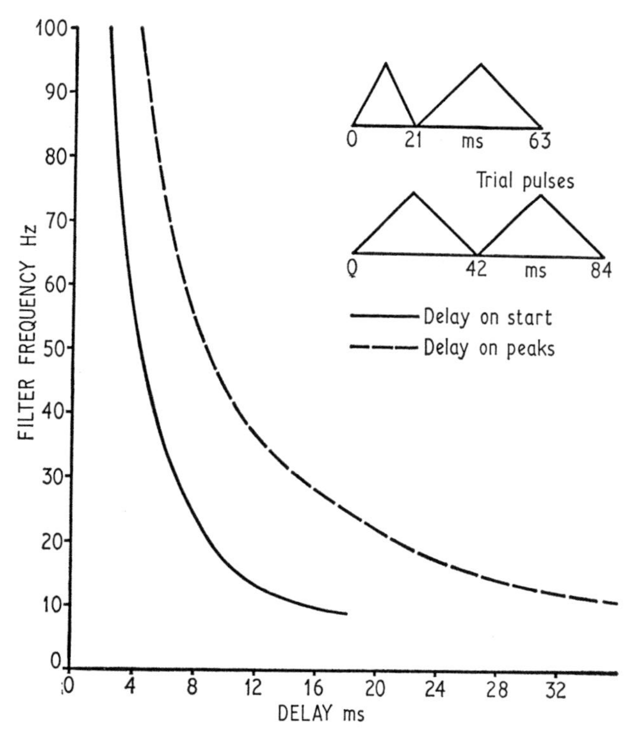

Fig. 1.4. The effects of delays through a typical variable filter with time-bases similar to typical crash transients

It is standard practice to check that accelerometers are functioning correctly by integrating the area under the deceleration traces and equating it to the impact velocity and vehicle deformation. Only the part of the acceleration trace up to maximum deformation is considered and the second integration is usually accomplished by multiplying the area under the trace by the time to the centroid of this area.

The time to maximum vehicle deformation is obtained from the analysis of the high-speed cine film and it is essential, therefore, that real-time correlation be achieved between accelerometer and cine-film. This does not mean that filtering should not be used but that the delay characteristic of every instrument should be known, including the variation of a time-delay with frequency of the filter employed.

In the following discussion of vehicle impact tests any results given will conform to the following constraints:

(i) The level of filtering employed will be specified.

(ii) The area under body-mounted accelerometer traces will equal the impact velocity within ±5 per cent.

(iii) Accelerometer traces used for comparison purposes will have the same response characteristics.

Occupant simulation

For the various impact tests discussed in this paper the 50th percentile Sierra Stan anthropometric dummy was employed to simulate the vehicle occupant.

Whilst this dummy is one of the most sophisticated on the market it is not a perfect simulation of *homo sapiens*. This is particularly so with regard to the dynamic behaviour of the cervical spine. A detailed evaluation of the 95th percentile Sierra anthropometric dummy by J. A. Searle and C. M. Hazelgrove of MIRA demonstrated other unrepresentative features.

The Sierra range of dummies are, however, a considerable improvement over the previous generation of test dummies. Furthermore, this type is widely used by the automotive industry on an international basis and it is therefore possible to correlate test results from different centres.

All occupant displacement, deceleration and seat-belt load data given in this paper are related to the 73·5 kg 50th percentile Sierra Stan anthropometric dummy.

Occupant restraint

To ensure correlation between the various impact tests, a standard lap and diagonal seat belt was employed for all the restrained occupant tests referred to in this paper.

FRONTAL IMPACTS

Standard test method

The standard frontal impact test is the 50 km/h impact on a massive concrete block. This test is used as the basis of three Federal Motor Vehicle Safety Standards for the U.S.A.

(i) Steering wheel horizontal penetration (FMVSS 204).
(ii) Fuel tank integrity (FMVSS 302).
(iii) Windscreen retention (FMVSS 212).

Relationship between frontal-barrier and vehicle-to-vehicle impacts

What does the frontal barrier impact test mean in terms of the real accident situation? It is obviously a true simulation of an impact of a vehicle on a massive flat-faced object at 90° but how does it relate to a vehicle-to-vehicle impact? It is generally considered to be equivalent to a frontal impact between two identical vehicles, each travelling at 50 km/h.

The deformation and vehicle/occupant deceleration values for a 50 km/h vehicle-to-vehicle frontal impact between two identical vehicles are compared in Table 1.1 with similar values recorded for a 50 km/h frontal barrier impact on the same type of vehicle. There is undoubtedly a high degree of correlation between the results, confirming that the barrier impact is a reasonable simulation of the impact of a vehicle on an identical vehicle.

It is interesting to make the same comparison for unlike vehicles. Table 1.1 gives results obtained for two vehicles of significantly different masses. The occupant of the heavier car benefits considerably from the increased stopping distance obtained by driving the lighter vehicle backwards.

A vast number of tests would be necessary to determine all the factors which contribute to the impact performance of dissimilar vehicles. The deformation pattern which results is dictated, apart from the relative speeds and masses of the vehicles, by their relative widths which to some extent determine the contact between the hard and soft points on the respective vehicles. It is clear, therefore, that a barrier impact cannot simulate either the accelerations or the deformations of impacts between dissimilar vehicles.

The dynamics of vehicle-to-vehicle impacts agrees with simple theory. The velocity change of each vehicle is inversely proportional to the vehicle masses and can be obtained as follows:

Let M_1 = Mass of first car in kg
M_2 = Mass of second car in kg
V_1 = Velocity of first car on impact in m/s
V_2 = Velocity of second car on impact in m/s
V_3 = Velocity after impact of both vehicles in m/s

and, considering conservation of momentum,

$$M_1V_1+M_2V_2 = (M_1+M_2)V_3$$

$$V_3 = \frac{M_1V_1+M_2V_2}{M_1+M_2}$$

$$V_3 = \left(\frac{M_1}{M_1+M_2}\right)V_1+\left(\frac{M_2}{M_1+M_2}\right)V_2$$

$$V_3-V_2 = \left(\frac{M_1}{M_1+M_2}\right)V_1-\left(\frac{M_1}{M_1+M_2}\right)V_2$$

$$V_3-V_2 = \left(\frac{M_1}{M_1+M_2}\right)(V_1-V_2)$$

Table 1.1. Comparison of results for vehicle-to-vehicle, and vehicle-to-barrier impacts

Type of test	Test vehicle	Velocity	Peak 'g' levels (100 Hz filter)				Maximum seat belt loads, kN			
			'*B*' pillar	Time, ms	Dummy head	Time, ms	Diagonal	Inner lap	Outer lap	Time, ms
Vehicle-to-vehicle	Type (1)—Medium	50 km/h	28	62	61	83	7·7	7·9	3·0	87
	Type (1)—Medium	50 km/h	29	61	60	100	7·6	9·8	3·4	98
Vehicle-to-barrier	Type (1)—Medium	50 km/h	27	68	70	94	8·9	9·1	3·0	90
Vehicle-to-vehicle	Type (2)—Heavy	50 km/h	26	52	48	121	—	—	5·4	81
	Type (3)—Light	50 km/h	48	38	110	86	10·85	11·7	5·9	64
Vehicle-to-barrier	Type (2)—Heavy	50 km/h	30	62	31	124	7·0	11·6	1·8	89
	Type (3)—Light	50 km/h	44	17	142	66	8·4	9·0	1·5	67

Change in velocity of car of mass M_2

$$= \left(\frac{M_1}{M_1+M_2}\right)(V_1-V_2)$$

In the case of unlike vehicles, covered in Table 1.1, the calculated velocity changes of the heavier and lighter cars are 37 km/h and 63 km/h, respectively. Referring to the actual vehicle deceleration traces and integrating, velocities of 42 km/h and 68 km/h are obtained which, after subtraction of the rebound velocity (coefficient of restitution, $e = 0{\cdot}09$) become 38 km/h and 62 km/h. This, therefore, confirms the linearity between velocity change and mass.

The relationship between impact velocity and deformation

The majority of vehicle-to-barrier tests have been conducted at 50 km/h to ensure compatibility with safety legislation. What, however, is the relationship between vehicle deformation/deceleration levels and the velocity of impact? To investigate this, frontal-barrier impacts were carried out at 30 km/h, 50 km/h, 70 km/h and 80 km/h on four nominally identical vehicles. It was immediately apparent from the results that the relationship between velocity and deformation was linear between 30 and 70 km/h. Furthermore, for all impact speeds the time to maximum deformation was similar (80 ± 5 ms) (Fig. 1.5).

This indicates that, in the 30–70 km/h speed range, a square law exists between vehicle dynamic stiffness and deformation. The result at 80 km/h suggests that at this speed the square law no longer applies. This could be due to the dynamic stiffness of the vehicle becoming very high at a time its main mass has already been brought to rest. Further tests at 70 km/h and 80 km/h would be necessary, however, to prove this hypothesis.

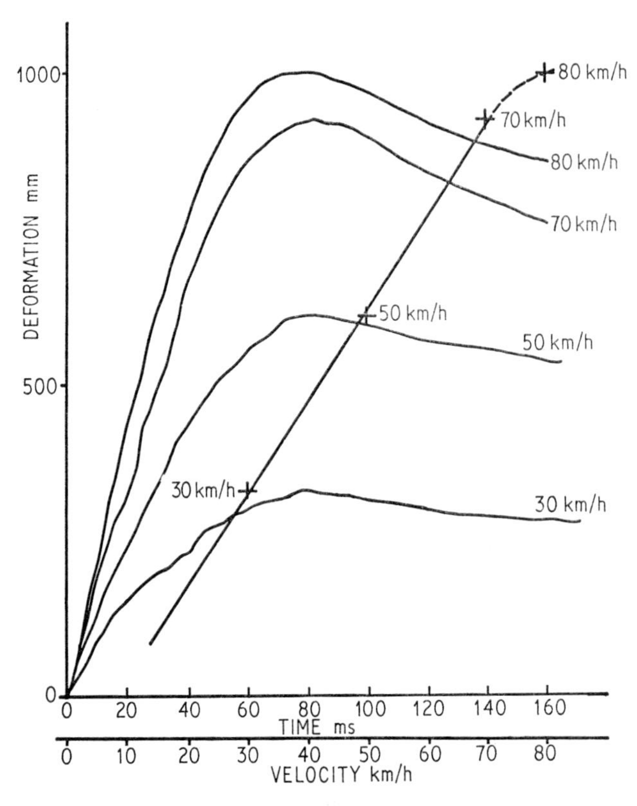

Fig. 1.5. The effect of barrier impact speed on deformation of a standard test vehicle

The total lap and diagonal seat-belt loads recorded for the 30, 50 and 70 km/h barrier impacts were 10 kN, 20 kN and 32 kN, respectively. At 70 km/h the loading would have been higher but was limited by the impact of the occupant's head on the structure. The severity index

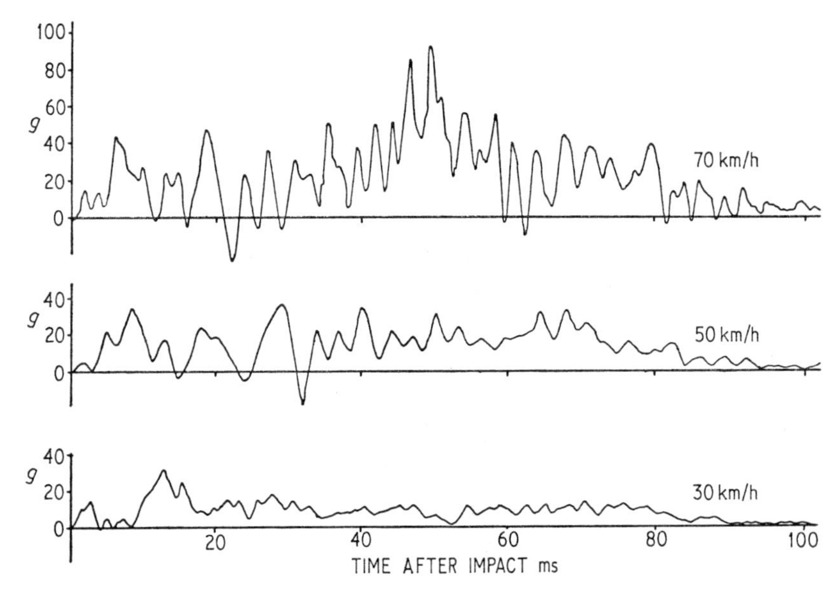

Fig. 1.6. Deceleration traces of a standard test vehicle hitting the barrier at three different speeds

(S.I. = $\int_{t_0}^{t} a^{2\cdot5}\,dt$) for this impact of the dummy's head was 4700 and would therefore have resulted in death.

The deceleration/time plots at the base of the '*B*' post for the above impact speeds are compared in Fig. 1.6. The pulse durations are similar for each of the impact speeds as was previously noted for the times to maximum deformation. It is these times that would have to be increased if the impact performance of a vehicle, were to be improved as regards structural deceleration.

Pursuing this point further, Fig. 1.7 gives the deformation times for a number of different vehicles: the tendency towards a mean about 80 ms is again evident. The vehicles which reach maximum deformation after 65 ms and 110 ms, shown on the graph, significantly correspond to the lightest and the heaviest vehicles tested, respectively.

Effect of vehicle mass on barrier impact performance

Whilst the effect of mass is demonstrable for vehicle-to-vehicle impacts, the position is not as clear for the barrier impacts. Two factors make their investigation difficult:

(i) Differentiating between stiffness and mass effects.
(ii) Test scatter.

It is possible to overcome (i) by confining tests to one type of vehicle. Fig. 1.8 gives the deformation values recorded

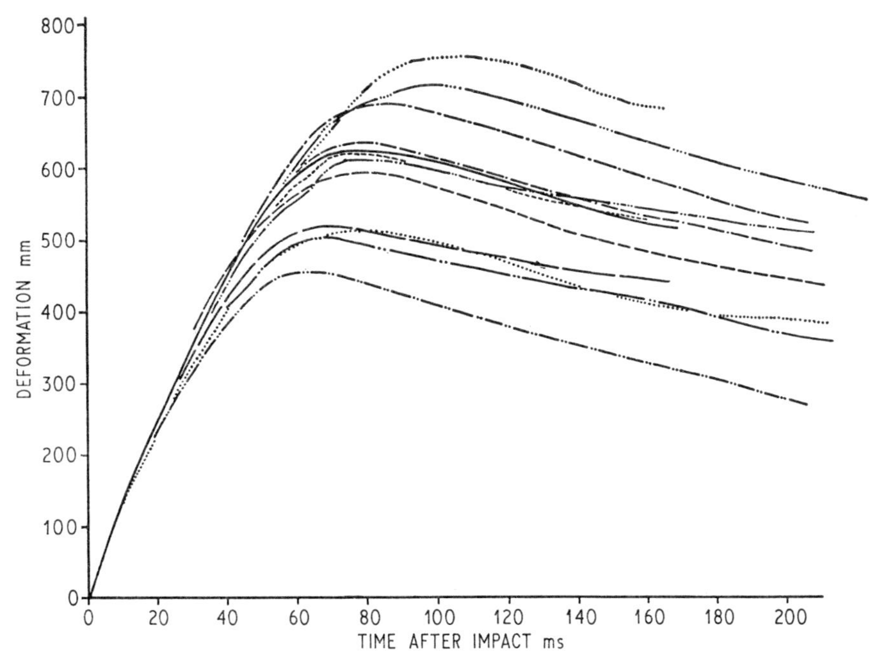

Fig. 1.7. A number of barrier impacts at 50 km/h show a tendency towards a mean value of 80 ms

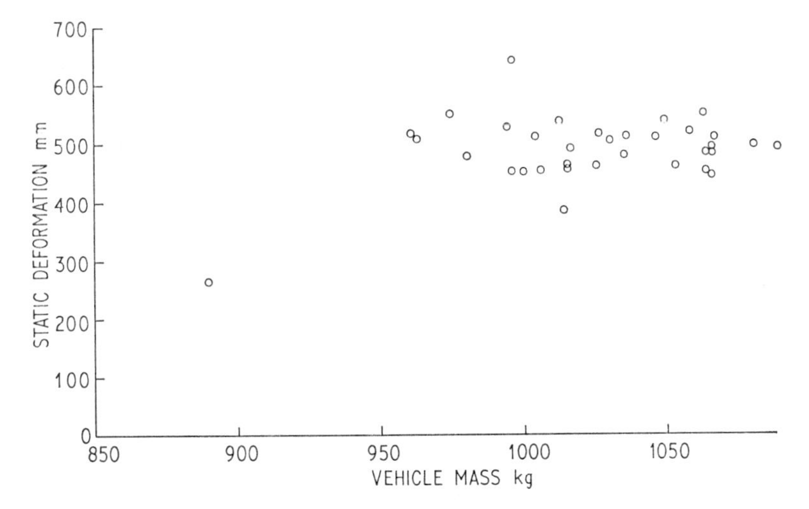

Fig. 1.8. Static deformations for various weights of vehicle of the same model

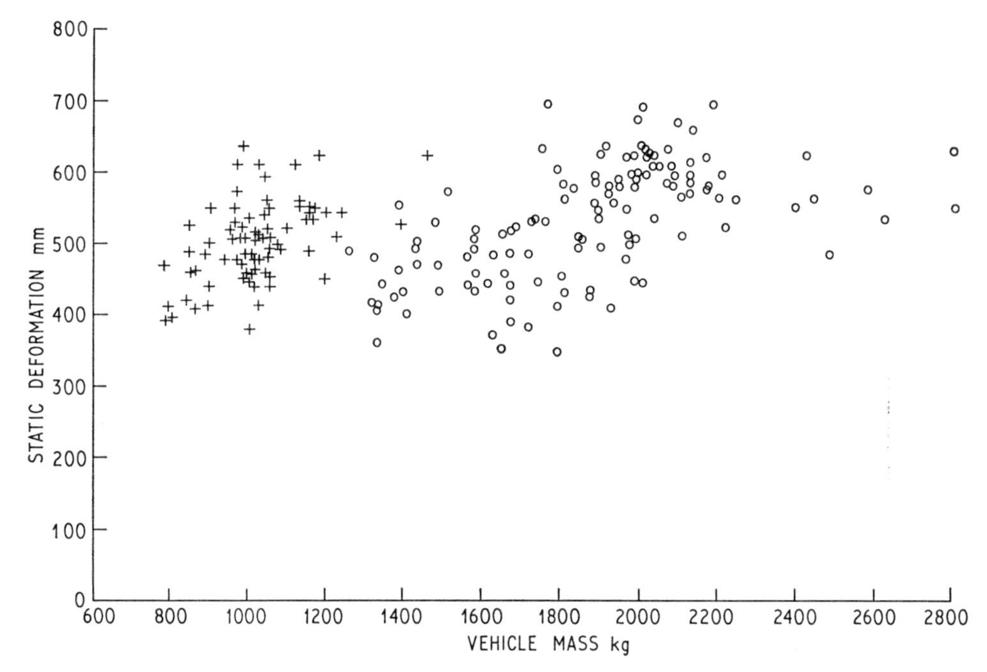

Fig. 1.9. Static deformation against mass for various types of car

for 34 barrier impacts at 50 km/h on one type of vehicle over a range of test weights. This figure effectively illustrates the degree of scatter in this type of test. It is difficult, over the weight range, to establish any accurate relationship between mass and deformation.

A similar picture emerges when plotting deformation against vehicle mass over a much wider weight range. Fig. 1.9 gives the corrected 50 km/h barrier-impact deformation values for a large number of tests on a wide range of European and American vehicles.

The unrestrained occupant

The structural deceleration characteristics of the vehicle are only important if the occupant is attached to that structure. The unrestrained occupant cannot benefit from 'ride-down' with the vehicle structure. If a vehicle is travelling at 50 km/h immediately prior to impact then the unrestrained occupant maintains this 50 km/h velocity until brought to rest against the now stationary internal structure.

The effect of friction between the occupant and his seat is negligible. This is illustrated in Fig. 1.10 which compares, on a time basis, vehicle deformation with occupant displacement for a standard 50 km/h frontal barrier impact. The slope of the occupants displacement/time curve remains constant until maximum vehicle deformation has occurred.

REAR IMPACTS

Standard test methods

There are three main types of in-line rear impact test:

(i) Test vehicle (target) against S.A.E. 1815 kg mobile barrier (bullet) at 32 km/h.

(ii) Test vehicle propelled backwards into a 100 tonne static block at 25 km/h.

(iii) Test vehicle against similar vehicle travelling at 50 km/h.

Methods (i) and (ii) form the basis of F.M.V.S.S. and Economic Commission for Europe proposals for safety regulations related to fuel-tank integrity. The rationale given by F.M.V.S.S. for adopting the flat 2·7 m wide, 1815 kg mobile barrier for rear impact tests was that it provided a reproducible test result and created damage to the rear of a passenger car at 32 km/h, similar to that resulting from a vehicle-to-vehicle test at 50 km/h (case iii).

The 25 km/h rear impact with the 100 tonne static

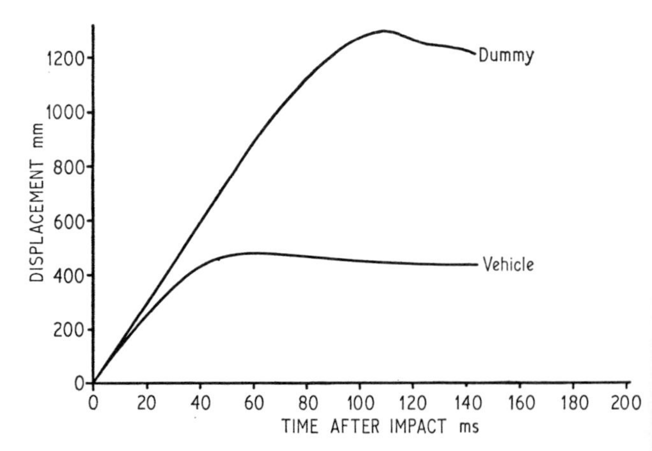

Fig. 1.10. Displacement of vehicle and of an unrestrained dummy in a 50 km/h frontal barrier impact

barrier is also intended to simulate the 50 km/h vehicle-to-vehicle impact (case iii).

Comparison of test methods

To assess how realistic the above rear-impact simulations are, a series of tests on nominally identical vehicles were made to compare the three test methods (i), (ii) and (iii).

Before discussing the test results, however, it is worthwhile to make the comparison on a theoretical basis.

For the mobile barrier impact the conservation of momentum gives:

$$M_1V_1+M_2V_2 = (M_1+M_2)V_3$$

where V_1 = Test vehicle velocity in m/s

V_2 = Mobile barrier velocity in m/s

V_3 = Velocity at maximum deformation of test vehicle in m/s

M_1 = Test vehicle mass in kg

M_2 = Mobile barrier mass in kg

therefore
$$V_3 = \frac{M_1V_1+M_2V_2}{M_1+M_2} \quad . \quad . \quad . \quad (1.1)$$

Considering conservation of energy:

$$\tfrac{1}{2}M_1V_1^2+\tfrac{1}{2}M_2V_2^2 = \tfrac{1}{2}(M_1+M_2)V_3^2+\int F\,ds$$

Energy spent in deforming test vehicle therefore equals:

$$\int F\,ds = \tfrac{1}{2}(M_1V_1^2+M_2V_2^2-M_1V_3^2-M_2V_3^2) \quad . \quad . \quad . \quad (1.2)$$

With regard to the static barrier impact, 'conservation of energy' gives:

$$\tfrac{1}{2}M_1V_{1B}^2 = \int F\,ds \quad . \quad . \quad . \quad (1.3)$$

To obtain the same deformation in a static barrier impact as from a mobile impact, F ds must be the same for both. Hence:

$$\tfrac{1}{2}M_1V^2 = \tfrac{1}{2}\langle M_1V_1^2+M_2V_2^2-(M_1+M_2)V_3^2\rangle$$

$$V^2 = V_1^2+\frac{M_2V_2^2}{M_1}-\left(\frac{M_1+M_2}{M_1}\right)V_3^2 \quad . \quad (1.4)$$

Substituting for V_3 from (1.1) gives:

$$V^2 = V_1^2+\frac{M_2V_2^2}{M_1}-\left(\frac{M_1+M_2}{M_1}\right)\left(\frac{M_1V_1+M_2V_2}{M_1+M_2}\right)^2$$

Now for mobile barrier impact $V_1 = 0$

$$V^2 = \frac{M_2V_2^2}{M_1}-\frac{M_2^2V_2^2}{M_1(M_1+M_2)}$$

$$\frac{V}{V_2} = \sqrt{\frac{M_2}{M_1}-\frac{M_2^2}{(M_1+M_2)}M_1} \quad . \quad (1.5)$$

Equation (1.5) can be stated in non-dimensional terms by letting

$$\frac{M_2}{M_1} = M_R, \quad \frac{V}{V_2} = V_R$$

Thus the ratio of static to mobile impact velocity becomes:

$$V_R = \sqrt{\frac{M_R}{1+M_R}} \quad . \quad . \quad . \quad (1.6)$$

By a similar process it can be shown that if both the mobile barrier and the test vehicle moving at the same speed, impact head-on the equation (1.6) becomes:

$$V_R = 2\sqrt{\frac{M_R}{1+M_R}}$$

where V_R is the ratio between impact velocities of the test vehicle when hitting a static block or a mobile barrier, respectively.

For a standard 32 km/h S.A.E. 1815 kg mobile barrier rear-impact with a test vehicle of mass 1067 kg,

$$V_R = \sqrt{\frac{1{\cdot}70}{1+1{\cdot}70}} = 0{\cdot}8$$

The static impact speed equivalent to the 32 km/h would equal $0{\cdot}8\times 32 = 25{\cdot}6$ km/h.

Table 1.2 compares the results of a 32·5 km/h mobile barrier impact, a 22·9 km/h static impact and a 51·5 km/h vehicle-to-vehicle impact. Identical test vehicles, weighing 1067 kg, were used for all the tests.

It is evident from the table that the structural collapse characteristics obtained for the mobile barrier and static barrier impacts are similar. The damage to the rear side-members and fuel tank was virtually identical. The structural decelerations are also similar, particularly if it is considered that the static barrier speed was 12 per cent below the theoretical equivalent to the mobile-barrier impact speed. Comparison of vehicle-to-vehicle results

Table 1.2. Comparison of test results

Target	Bullet	Spd. km/h	Target acc: peak 'g'	Front dummy long. chest 'g'	Rear dummy long. chest 'g'	Maximum static deformation mm
Vehicle 'A'	Vehicle 'A'	51·5	9	—	—	660*
Vehicle 'A'	Mobile barrier	32·5	11	4·9	18·5	180
Vehicle 'A'	Static barrier	22·9	8·5	—	25·6	178
Vehicle 'B'	Vehicle 'B'	51·3	11	—	30·2	580*
Vehicle 'C'	Mobile barrier	33·7	13	8·3	22·4	274
Vehicle 'D'	Mobile barrier	33·5	13	5·4	—	183
Vehicle 'E'	Mobile barrier	38·9	14	6·4	73·8	171

* Note: Deformation of target vehicle is irregular in vehicle/vehicle impacts.

with those from the barrier impacts shows good agreement for structural deceleration levels.

This is not, however, the case with vehicle deformation patterns. The barriers offer a flat, unyielding surface which spreads the impact force evenly over the entire rear surface. With vehicle-to-vehicle impact the deformation is greater and more localized. The hard central mass of the impacting vehicle (engine and sidemembers) easily penetrates the soft area between the rear sidemembers of the target vehicle.

Further tests comparing vehicle-to-vehicle with static and mobile barrier impacts on other types of vehicle endorsed these results and enabled the following conclusions to be drawn:

(i) The static barrier can be interchanged with the mobile barrier for the study of structural collapse, including fuel-tank integrity. Use of the static barrier with its comprehensive lighting and camera facilities would, in general, permit better data retrieval.

(ii) Barrier impacts provide a reliable and repeatable method of simulating the decelerations experienced by a vehicle structure and its occupants in an in-line vehicle-to-vehicle rear impact. They do not, however, effectively simulate the rear end deformation characteristics.

Use of the barrier impact for optimization of seat/head restraint design

The occupant decelerations recorded for the various rear impacts tested are not in themselves unduly severe. It is immediately apparent, however, from analysis of high-speed cine films, that they are sufficient to result in serious 'whiplash' effects, causing hyper-extension of the cervical spine.

It is here that the barrier impact serves a useful purpose in providing a repeatable test which can be employed to study and compare occupant kinematics and develop optimum seat/head restraint designs.

From the decelerations given in Table 1.2, it can be seen that the values for the front-seat occupant are significantly lower than those of the rear passenger. This is basically due to the greater distance over which the front occupant is decelerated. The majority of current European front seat backs deform under the inertia loading generated by the dummy under impact test conditions, whereas rear seat backs are relatively unyielding. The peak deceleration for the front-seat occupant occurs after approximately 165 ms. This is the time during which the head has either rolled over the seat-back or has impacted on the rear seat. Before this there is little horizontal head acceleration and then, as the seat-back yields, the body rotates backwards which prevents relative motion between head and trunk. The vertical head acceleration becomes apparent at 45 ms and increases to a maximum when the head reaches a horizontal position.

Therefore it appears from the barrier impact tests that, for front seats without head restraints, a yielding seat back would be advantageous. However, there are inherent disadvantages in a seat with a yielding back which, the author feels, make them a poor second choice compared with the strong seat and integral head restraint. These are:

(i) Yield of seat-back can only be optimized for one combination of occupant weight, size and impact speed.

(ii) For high-speed impacts the inclination of the seat-back may act as a 'launching ramp' for the occupant, causing severe injuries due to secondary impact.

(iii) The yielding seat back could be injurious to rear-seat occupants, particularly in smaller vehicles.

SIDE IMPACTS

Standard test methods

With the exception of a rear, side, 90° mobile barrier impact to assess fuel-tank integrity, there are no dynamic tests to form the basis of safety legislation. This is due, mainly, to the extreme difficulty in obtaining clear and repeatable data from the complex motions involved in this type of impact.

For this reason the F.M.V.S.S. proposal for measuring side-door strength is related to a static test. The latter test consists in crushing the door, representatively mounted in a 'body-in-white', by means of a vertical steel semi-cylinder. This type of test undoubtedly overcomes the measurement and repeatability problems associated with the dynamic test and is probably a much sounder basis for legislation at present.

Dynamic test results

Dynamic tests are essential for development work as they remain the only means of investigating the structure of the vehicle side in relation to occupant motion. The dynamic test, however, can only simulate specific cases of the numerous types of accident that occur. It is sensible, therefore, that the most severe type of side impact should be chosen—that of the 90° collision.

The most widely used dynamic tests which are intended to simulate this 90° side impact are described below.

The S.A.E. mobile barrier

The mobile barrier is employed as the 'bullet' vehicle to remove one of the test variables, that of 'bullet' car deformation. This simplification, however, is achieved at the expense of an unrealistic target car deformation pattern.

Fig. 1.11*a* and *b* show the static deformation patterns recorded on the same type of test vehicle for the side-impact tests listed in Table 1.3. Results for the 45° impacts from the rear are not given. In all cases this latter type of impact is less severe, from a deformation point of view, than the 45° from the front and the straight 90° impacts.

It is evident from Fig. 1.11*a* and *b* that, in the vertical profile plane, there is a lack of correlation between vehicle and mobile barrier deformation patterns. Closer agreement occurs in the lateral profile plane, as shown in Fig.

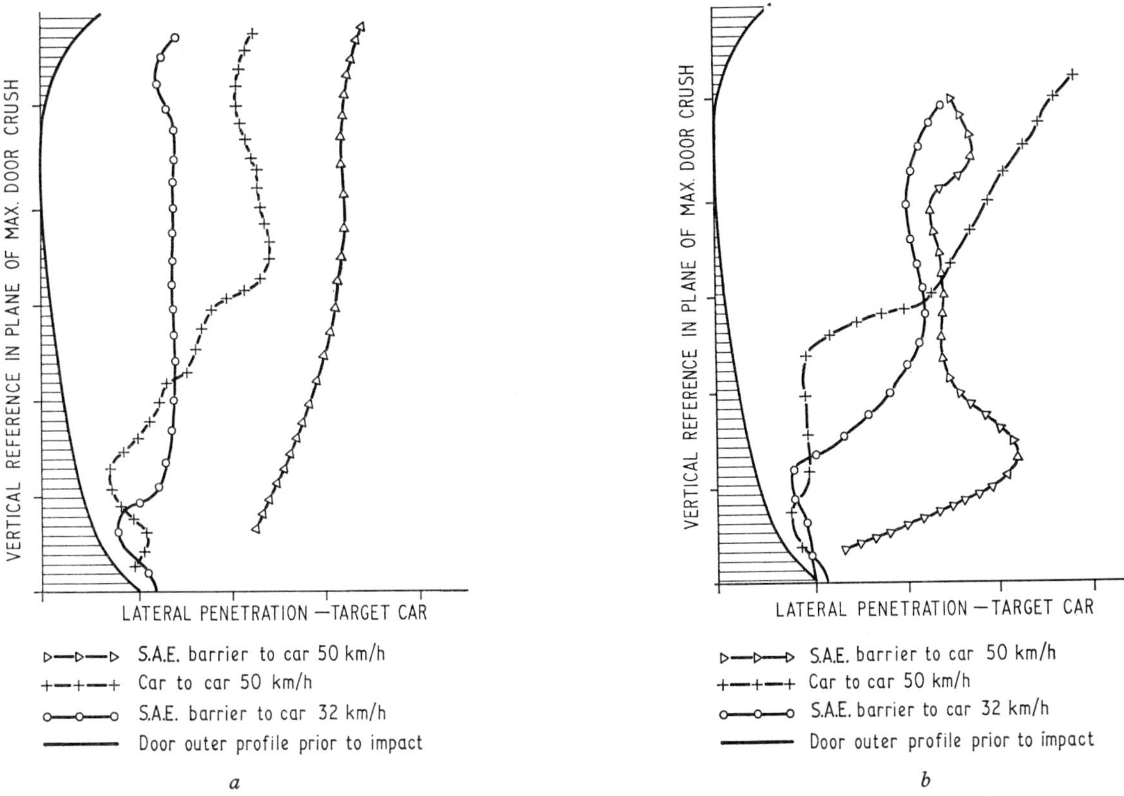

Fig. 1.11. Side penetrations in car-to-car, 90° side-impacts and S.A.E. barrier test (*a*) and 45° side impacts (*b*)

1.12. Contour modifications to the face of the S.A.E. mobile barrier are being carried out in the U.S.A., in an attempt to improve the correlation between barrier and car-to-car impacts. Until this work is satisfactorily completed, use of the mobile barrier for investigating side impact performance is severely limited.

Vehicle-to-vehicle side impacts

The side impact content of the Ford research programme has concentrated on the 50 km/h, 90° and 45° from the front, vehicle-to-vehicle impacts.

A major problem with this type of test is the accurate measurement of dynamic penetration. With suitably located high-speed cine-cameras it is possible to obtain the overall dynamic deformations of the target and bullet vehicles. It is also possible to measure the final static deformation of the two vehicles. With this information, it is possible to obtain a reasonable estimate of the actual dynamic penetration of the target car. Fig. 1.13 gives an actual displacement analysis for a typical 90° side impact of two nominally identical vehicles. The value of this analysis will become apparent in the later discussion on the effects of side stiffening.

Deformation patterns

From the severity point of view, it is difficult to differentiate between the 45° from the front and the 90° side impacts. Both tests are extremely severe and the resulting deformation patterns are mainly determined by the following factors:

(i) Strength of the door hinges and attachments.

(ii) Strength of the 'B' post, header rail and sill attachments.

Table 1.3. Side impact tests

Target	Bullet	Direction*	Speed of bullet
Stationary test vehicle	SAE mobile barrier	90°	32 km/h & 50 km/h
Stationary test vehicle	SAE mobile barrier	45° from rear	32 km/h & 50 km/h
Stationary test vehicle	SAE mobile barrier	45° from front	32 km/h & 50 km/h
Stationary test vehicle	Test vehicle	90°	50 km/h
Stationary test vehicle	Test vehicle	45° from rear	50 km/h
Stationary test vehicle	Test vehicle	45° from front	50 km/h

* For the 90° impact the centre lines of the bullet vehicles and the target car door are aligned. For the 45° impact the bullet vehicle is aligned to make initial contact with the target car door opening line.

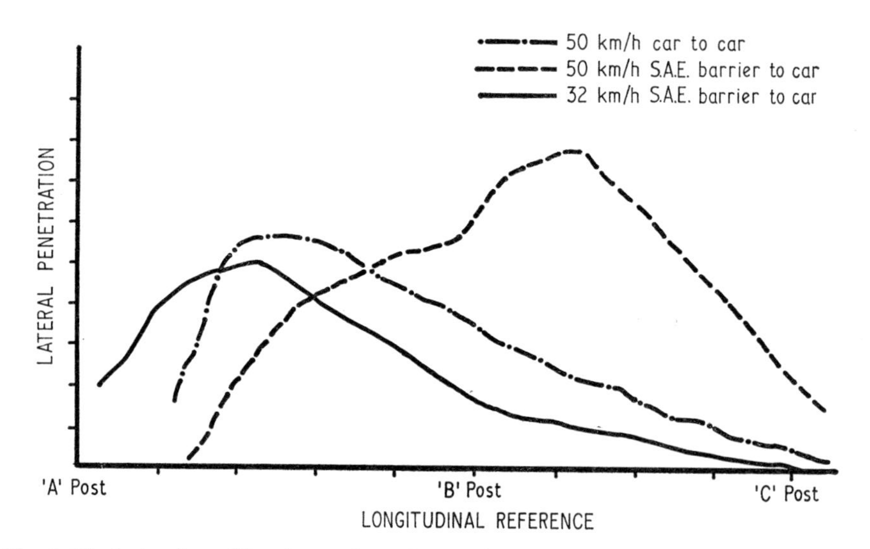

Fig. 1.12. Lateral profile plane of maximum door penetration in 45° front impact

(iii) The height of the sill member relative to rotational axis of bullet car road wheels.

Factors (i) and (ii) are obvious, but (iii) requires amplification. A low sill member on the target car enables the bullet vehicle to literally 'ride up' the side of the door. This 'ride up' does not necessarily cause greater horizontal penetration of the target car but occupants on the impacted side of the vehicle would undoubtedly sustain severe crushing injuries to the lower parts of the body.

Damage to the bullet vehicle in all cases is of a minor nature, and is confined to the front 150 mm of structure. There is normally little evidence of damage to the front side members. In terms of the total kinetic energy to be dissipated, it is estimated that 10–15 per cent goes into deformation of the bullet vehicle.

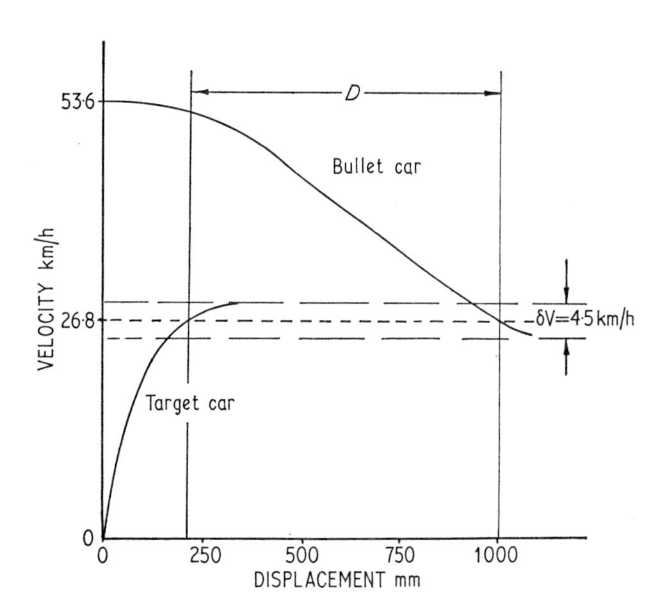

Fig. 1.13. Displacement of cars in 90° side-impact at various speeds

Structural and occupant accelerations

It can be seen from Fig. 1.14 that the structural acceleration levels recorded from the 50 km/h 45° or 90° side impact are lower than those which occur for the 50 km/h frontal barrier impact. This is not so with regard to occupant acceleration. Table 1.4 show that values in excess of 100*g* can be recorded for the head of the front-seat occupant on the side of the vehicle which is hit. High-speed cine-film analysis indicates that these high values occur when the head impacts either the door in the area of the glass aperture or the hood top of the intruding bullet vehicle. In all cases the door glass fractures on initial impact of the bullet vehicle and only debris remains by the time head impact occurs.

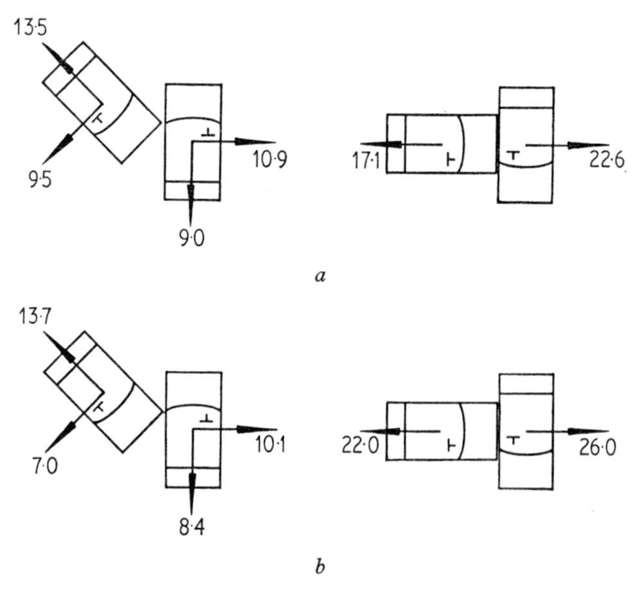

Fig. 1.14. Peak '*g*' levels for 45° and 90° side-impacts, (*a*) averaged over several models and (*b*) typical results for a 1100 kg vehicle

Table 1.4. Peak acceleration values recorded for dummies heads in side impacts

Test	Vehicle type 1			Vehicle type 2			Vehicle type 3		
	Lat.	Vert.	Long.	Lat.	Vert.	Long.	Lat.	Vert.	Long.
90° side impacts Front occupant impact side	110g 64 ms	29g 37 ms	43g 49 ms						
Front occupant far side				25g 133 ms	60g 134 ms	25g 132 ms	10g 177 ms	28g 151 ms	—
45° side impacts Front occupant impact side	105g 70 ms	67g 72 ms	134g 70 ms	88g 78 ms	63g 75 ms	79g 96 ms	36g 96 ms	61g 72 ms	91g 96 ms

N.B. Filter level 100 Hz.

Side stiffening of body structure

The aim of current proposals for safety legislation conserning side intrusion resistance appears to be the progressive stiffening of the body side-structure. Is this really what is required for occupant protection? Before considering the answer to this controversial question, it is worthwhile to examine the basic velocity/displacement relationships which determine the severity of the occupant's impact with the interior structure.

Fig. 1.15 illustrates, in diagrammatic form, the relationship between the velocity of impact and occupant position for a 50 km/h 90° side impact between vehicles of similar weight. The figure also indicates the effect increased stiffening would have on this velocity/displacement relationship and enables the following conclusions to be drawn:

(1) Increasing the stiffness of the side of a body accelerates the target car to its final speed more rapidly.

(2) Increasing the stiffness of the side of a body redirects more of the energy to be dissipated into bullet car deformation.

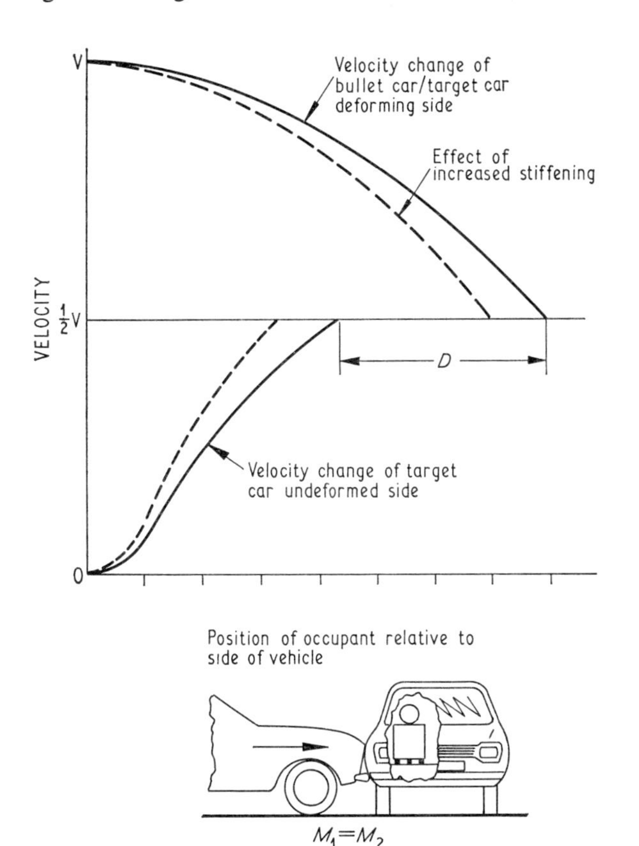

Fig. 1.15. Occupant positions at various impact velocities, showing effect of side stiffening

Test results show that seat friction is negligible and that the unrestrained occupant remains effectively stationary in space until hit by the interior structure of the vehicle. From Fig. 1.15 it can be seen that the severity of this impact depends on the occupant's position. The effect of additional side stiffening is as follows:

(i) Occupants seated close enough to be hit by the deforming side structure would be better off if side stiffening were fitted.

(ii) Occupants seated far enough away not to hit the structure before a common velocity is reached by the colliding vehicles will neither gain nor lose from the addition of side stiffening.

(iii) Occupants seated close enough to non-deforming parts of the structure to hit it, prior to the common velocity of the vehicles being reached, would be worse off if side stiffening were fitted.

The adverse effect of (iii) is largely outweighed by the beneficial effects of (i). Furthermore, reference to Fig. 1.13 will show that, to benefit significantly from 'riding up' with the non-deforming structure, the rear-seat occupant of the target car in this test would need to be within 100 mm of the vehicle side-structure at the time of impact. It should also be noted that (iii) applies only for 90° impacts. Test results show that the occupant's relative motion is towards the point of impact and is influenced by the angle of impact. An occupant would reap greatest benefit from stiff side structure in a glancing blow type of impact (30°): only the advantageous factor (i) and (ii) would then apply.

There are obviously, however, practical and economic considerations which limit the effective door stiffness that can be achieved. Based on the 50 km/h 45° side impact test, a realistic aim would be a vehicle side strong enough to maintain structural continuity and sufficiently stiff to prevent crushing injuries to the occupants. Further occupant protection would then be more effectively provided by energy absorbing padding, rather than increased stiffening of the vehicle side.

ACKNOWLEDGEMENTS

The author wishes to thank the Ford Motor Company Limited for allowing him to publish this paper, and to express his appreciation of the support given to him by his colleagues in the Engineering Research Department.

APPENDIX 1.1

REFERENCES

(1) Emori, R. I. 'Analytical approach to automobile collisions', S.A.E. Paper 680016, 1968.

(2) Gadd, C. W. 'Use of weighted impulse criterion for estimating injury hazard', Paper 660793, 10th Stapp Car Crash Conference.

(3) Marquard, E. 'Fortschritte in der Berechnung von fahrzeug Zusammenstossen', *ATZ*, Vol. 68, March 1966.

(4) Kilbuszewski, J., Mackay, G., Fonseka, C., Blair, I. and Clayton, A. 'Cause and effects of road accidents', Vol. 1–5, University of Birmingham Road Accident Research Unit Report 1969.

(5) Grime, G. and Jones, I. 'Car collisions—The movement of cars and their occupants in accidents', Paper presented to I.Mech.E Automobile Division, January 1970.

(6) Searle, J. A. 'The optimization of occupant restraint for frontal impact', MIRA Report No. 1969–11, April 1969.

(7) Neilson, I. D. 'Simple representations of car and unrestrained occupant impacts in road accidents', RRL Report LR 249, 1969.

(8) Searle, J. A. and Hazelgrove, C. M. 'Anthropometric dummies for crash research', MIRA Bulletin No. 5, 1969.

(9) MacLean, R. B. and Berton, R. J. 'Potential applications for the movable barrier in vehicle crash testing', 1970 FISITA S.A.E. Paper No. 700408.

Paper 2

THE EFFECTS ON BODY STRUCTURES OF PRESENT AND POSSIBLE FUTURE SAFETY LEGISLATION, AND THE MATHEMATICAL SIMULATION OF BARRIER IMPACT

J. Curtis*

The wise body designer looks ahead to the type of statutory safety requirements that might be in existence by the time his current design has been on the road for some years. But the necessary testing of prototypes is expensive. The author describer a computer simulation of impact tests, based on empirical data from static crush tests and the equations of motion.

INTRODUCTION

BODY DESIGN ENGINEERS must consider many requirements in a new vehicle structure: for example, the static and dynamic stiffness of the complete body and resonant frequency of individual panels, weight, ease of manufacture and assembly, corrosion resistance and painting. Safety considerations are assuming more prominence, particularly since several countries have introduced legislation which has a bearing on the vehicle structure. The United States Federal Government has led the way with its Federal Motor Vehicle Safety Standards, and the Swedish Ministry of Road Safety (Statens Trafikaakerhetverk) has introduced safety regulations, some of which are based on the FMVSS. Furthermore, E.C.E., through Working Party 29, is well on the way to finalizing regulations which are likely to be generally adopted in Europe.

These regulations are continually subject to modification and enlargement, and it seems certain that those which affect body structural design will be made more stringent, and others introduced in future years.

Current Safety Regulations which affect body structures include the following.

Steering column horizontal penetration

FMVSS 204. This regulation limits horizontal rearward travel of the upper end of the steering column to 5 in of dynamic movement during a 30 mile/h barrier collision test.

F7 (Swedish). As FMVSS 204, but with a maximum dynamic rearward movement of the steering column of 127 mm. E.C.E. Reg. does not permit the use of a dummy.

Windshield retention

FMVSS 121. This regulation requires that not less than 75 per cent of the windshield periphery must remain attached to the vehicle after the 30 mile/h barrier collision test. It alternatively specifies that, when unrestrained 95th percentile adult male manikins occupy the outer front seats, not less than 50 per cent of the windshield periphery on each side of the vehicle centre line must remain attached to the vehicle after the 30 mile/h barrier collision test.

Fuel tank integrity

FMVSS 301. This regulation requires that not more than one ounce of fluid shall leak from the fuel tank during the 30 mile/h barrier collision test, and that subsequent leakage shall not exceed one ounce of fluid per minute.

F13 (Swedish). As FMVSS 301 but with 30 g substituted for one ounce.

Due to the considerable time between the initial design of a new body and its introduction, the body designer must try to predict the extent to which new or more stringent safety regulations will apply by the time the vehicle is introduced. He must also bear in mind that the

The MS. of this paper was received at the Institution on 24th June 1970 and accepted for publication on 30th June 1970.
* *Senior Project Engineer, Vauxhall Motors, Luton, Beds.*

model may run for several years with the same basic body and will be required to comply with the safety regulations throughout that time, without drastic modifications. It is important, therefore, to have at least a rough idea of what areas may be investigated for safety legislation, together with changes in the severity of existing regulations.

Potential future regulations

The following items are examples of this type of potential regulation:

(1) Impact on the rear of a stationary vehicle by a moving, 400 lb barrier at 20 mile/h, with the same standard of fuel retention as FMVSS 301.

(2) Possible raising of the test speed of the barrier frontal impact.

(3) Some limitations on the deceleration pulse in the barrier collision test might be imposed. For instance, some sort of system performance based on duration and magnitude of deceleration of the head or torso of the occupant, or a performance based on, say, floor-panel deceleration.

(4) Side impact. A static or dynamic test, with a maximum internal lateral deformation. If this test is dynamic, it is likely to be an impacting vehicle of a specified type striking the stationary test car at 45° just behind the front hinge pillar.

(5) Roll-over. This test could be static, but intended to simulate the damage caused by a dynamic roll-over accident. It could take the form of a specified load applied by a crusher rig to the front roof area, on one side of the vehicle, with the maximum deformation limited.

Mathematical models

On prototypes most of these tests are very expensive to carry out, and it is naturally desirable to reduce the amount of development testing which must be done. One possible way of doing this is to simulate the test with the aid of a mathematical model of the vehicle, and this has been attempted for barrier frontal impact.

General Motors Research have already had considerable success with this type of simulation. Their mathematical model has three degrees of freedom, representing the undeformed body, the engine and gearbox, and the conjunction of front engine-mounts, front of chassis frame, and rear of chassis frame (that is, from the front engine mount to the dash panel). The first two of these are assumed to have point masses.

These nodes are connected to each other and to the barrier by elements with known force/displacement relationships. The time-response of the three nodes can be determined by numerically integrating the equations of motion, from starting conditions of velocity into the barrier.

The force/displacement relationships for the various elements comprising the model are determined experimentally by crushing the appropriate parts of the vehicle structure in a specially designed rig. The tests are done slowly, and may be stopped at any time to investigate in detail the mode of deformation of the structural part being tested.

Originally, the model ignored damping effects in the elements, but these have since been included. They improve the accuracy of the simulation. The success of the G.M. Research methods encouraged us to attempt a similar programme, using three masses, instead of two.

A SIMPLE MODEL OF BARRIER IMPACT

The model to be described is a simple one with three point masses. A diagram of it is shown in Fig. 2.1.

The three masses represent:

(1) The undeformed part of the vehicle behind the front door pillars, including occupants, rear wheels and rear axle.

(2) The engine and gearbox unit.

(3) The front axle unit, including front unsprung masses, and the steering unit in the case of the vehicle analysed.

These masses are connected together by elements which produce forces dependent on their change in length. The computer programme finds these forces by linear interpolation of tables of force against contraction for each of the seven elements. The force/displacement curve of each element may thus have any shape.

The seven elements shown in Fig. 2.1 represent the following parts of the deformable front-end structure:

$F1$. The rear part of the front side members between the front suspension mounts and the non-deforming body. (This may include other items. For instance, on the Vauxhall *Viva*, the front suspension crossmember has rearward extensions, which are attached to the side members, and these are included in element $F1$.)

$F2$. The front part of the sidemember between the front suspension mounting and the front panel.

$F3$. The rear engine-mount plus the propeller shaft, rear-axle and suspension.

$F4$. Those items which directly connect the body to the barrier. That is, the front fender panels and the hood.

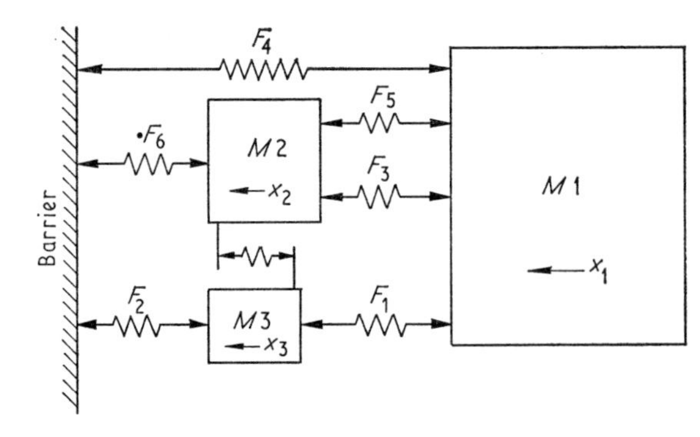

Fig. 2.1. A simple three-mass model of barrier impact

*F*5. The dash-panel and transmission tunnel. This element is normally zero at impact and may not engage until the relative displacement of masses 1 and 2 is several inches.

*F*6. The radiator, fan and sheet metal between the engine unit and the barrier. Again, this element usually has some 'slack' in it.

*F*7. The front engine-mounts.

Simplifying assumptions

It is assumed that rotation of the masses produces no significant changes in the crush characteristics of the elements. Damping of the elements is neglected, so that the shape of the force/displacement curves may be obtained from static tests carried out at low speed in the workshop. It is also assumed that interaction between the various elements is negligible, so that crushing of one has no effect on the crush characteristics of others.

It is assumed that various items at the front end of the vehicle, for instance, the front panel, lamps, battery and radiator, are brought instantly to rest on impact, without affecting the dynamics of the rest of the vehicle.

Equations of motion

Using these assumptions, the simultaneous equations of motion of the model may be written as follows:

$$M_1\ddot{x}_1 = -(F1 + F3 + F4 + F5)$$
$$M_2\ddot{x}_2 = F3 + F5 - F6 + F7$$
$$M_3\ddot{x}_3 = F1 - F2 - F7$$

with initial conditions

$$x_1 = x_2 = x_3 = 0$$
$$\dot{x}_1 = \dot{x}_2 = \dot{x}_3 = V$$

where V = test speed

A programme has been written for an IBM 360 computer to solve these equations and plot the resulting responses. The equations of motion are integrated, using a fixed-step 4th order Runge–Kutta routine, and the displacement, velocity and deceleration, together with the element forces, are printed for a given time interval.

In order to prevent the masses rebounding from the barrier during the simulation, the programme sets the acceleration of any mass to zero when its velocity reaches zero. This simple procedure could be refined by introducing the residual elastic rate of each element when its crush reaches a maximum.

When the programme was written it was hoped to verify it by testing parts of the vehicle on a static crusher, and thus obtaining the required empirical force/displacement curves of the elements. These tests must obviously be done with care to ensure that only horizontal motion of the centre of gravity of the mass is used. Similarly, the measured forces must be those producing horizontal forces on the appropriate masses.

Unfortunately, the hardware for these tests is not yet available, except in the United States, and consequently programme runs so far have been done only with estimated data.

DECELERATION

The results of one such simulation compared with test results for a Vauxhall *Viva* station wagon are shown in Figs 2.2, 2.3 and 2.4. The element force/displacement

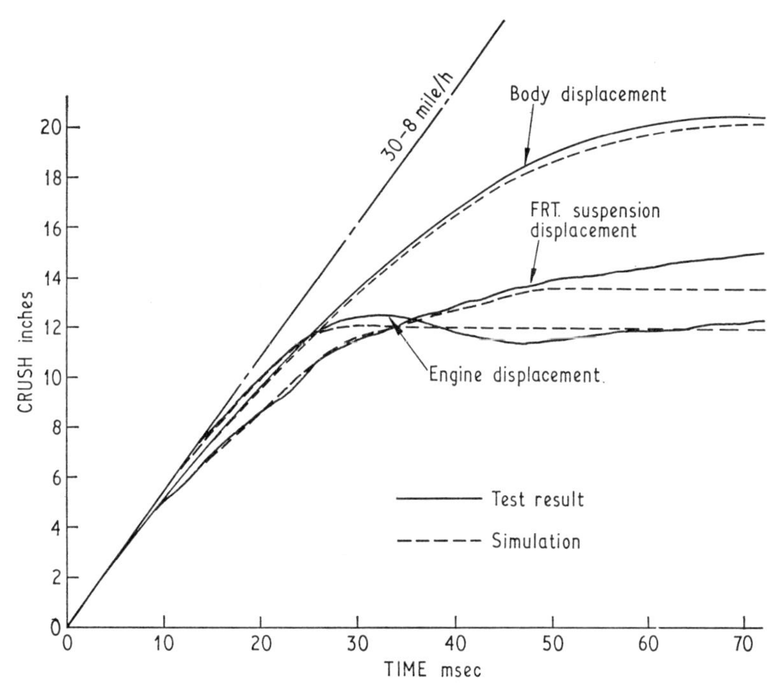

Fig. 2.2. Comparison of displacement with time, calculated on the basis of the model, and measured in a test

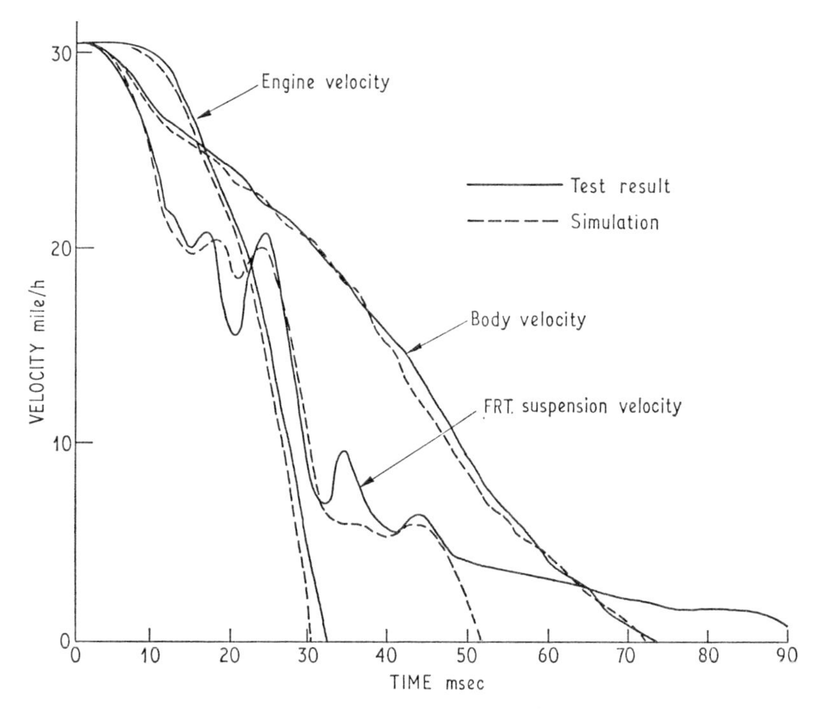

Fig. 2.3. Comparison of stopping characteristics of car components calculated on the basis of the model, and measured in a test

curves were arrived at by trial and error and used to predict the effects of barrier collisions at 15 and 40 mile/h, shown in Fig. 2.5.

If the deceleration pulse can be predicted with reasonable accuracy and the vehicle modified so as to obtain a desirable pulse the problem remains: what is a desirable pulse? There are two main candidates, the square-wave and the triangular wave. To get some idea of their relative merits we can take two simple examples.

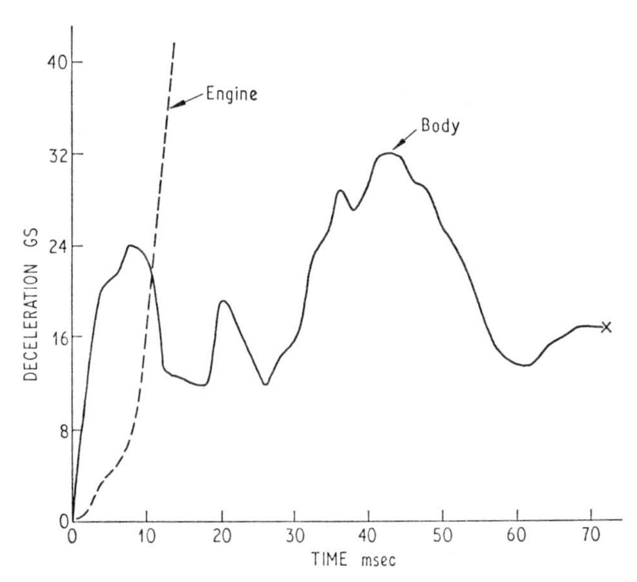

Fig. 2.4. Deceleration characteristics of engine and body

The first is a pure square wave with deceleration constant at $20g$ or 644 ft/sec^2. This type, shown in Fig. 2.6, gives the minimum peak deceleration for a given crush distance. Crush distance is likely to be the limiting parameter in practical design, and one definition of the 'best' pulse, therefore, could be one which subjects four totally restrained occupants to the minimum deceleration within a particular crush distance.

It should be clearly understood that there is an important difference between restrained and unrestrained occupants. If the unrestrained occupant is, say, two feet from the instrument panel at the moment of impact at 30 mile/h he will strike the instrument panel about 0·082 s after barrier impact when the vehicle interior is moving at less than one mile/h (assuming a vehicle crush of 19 in). The velocity of the second collision is about 29 mile/h and, under these circumstances, with a limited amount of crush available, the deceleration pulse shape makes little difference to the second collision velocity.

If the occupant is restrained, by seat belts for example, or if the free space in front of him is very small, say two inches, he will strike the interior, or take the slack from the seat belts about 0·024 s after barrier impact when the vehicle interior velocity is about 19 mile/h, with a second collision velocity of 11 mile/h. He then rides down the remainder of the vehicle deceleration pulse, with a deceleration depending on the vehicle deceleration, his own mass and the force/displacement characteristics of the restraint system.

Neilson (1)* gives a simple 'Injury Index' as a measure

* *References are given in Appendix 2.1.*

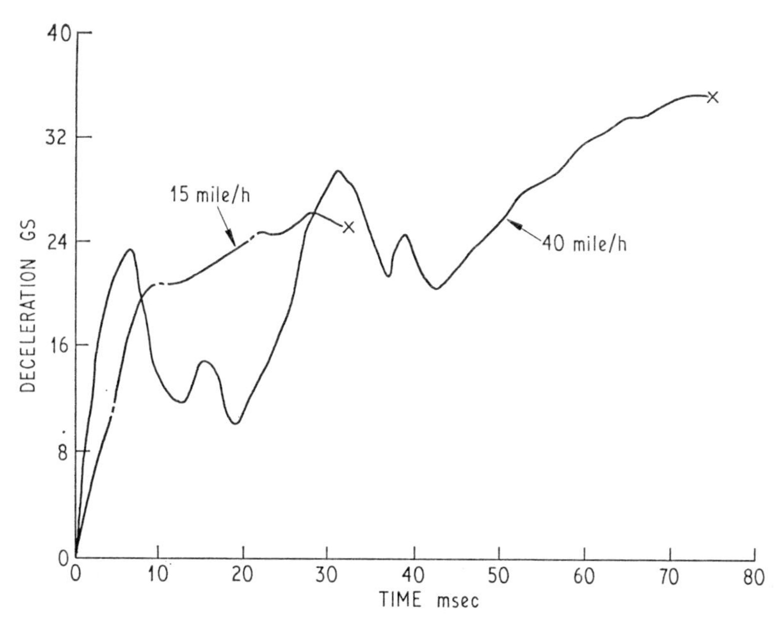

Fig. 2.5. Deceleration characteristics at two different speeds predicted from trial and error force/displacement curves

the severity of deceleration. This he defines as the ratio of the impact deceleration and the 'threshold deceleration' for a particular duration of deceleration. This threshold deceleration, above which serious injury could result, is arbitrary to some extent, and Neilson presents a suggested threshold curve in his report. This curve shows that tolerance to deceleration decreases as the duration of the deceleration increases, and both these factors must be considered when the severity of deceleration is discussed.

Clearly, the occupant experiences a different deceleration/time history than the vehicle structure round him, the difference depending on the occupant's mass, the force/displacement characteristics of the occupant restraint system, and the degree of slack or free space in it. In order to compare different types of structure deceleration pulse the Gadd Severity Index is used (**4**), with a weighting factor of 2·5.

Returning to the square wave example, the duration of the 20*g* deceleration is found to be 0·0682 s for an impact speed of 30 mile/h. The crush distance is found to be 18 in and the Severity Indcx 122.

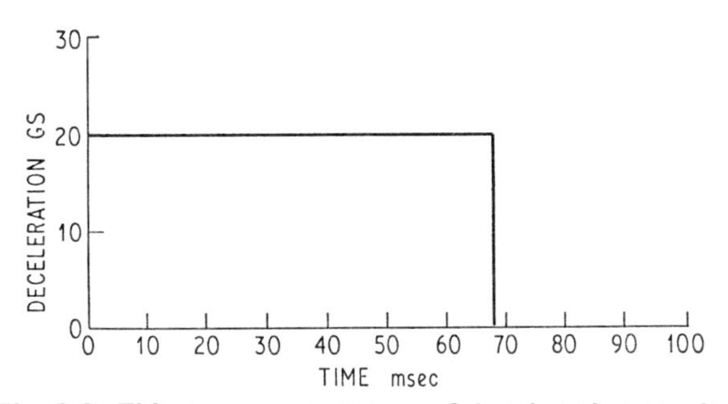

Fig. 2.6. This square wave type of deceleration results in the lowest 'peak' for a given crush distance

If we now turn to the triangular pulse, shown in Fig. 2.7, using the same impact time as with the square wave, we see that the peak deceleration is 40*g* and the crush distance is 12 in. The Severity Index for this pulse rises to 197. If the crush distance is now increased to 18 in to match the square wave, we find that the impact time increases to 0·102 s and the peak deceleration falls to 863 ft/s^2 or 26·8*g*. The Severity Index is reduced to 108, indicating a more tolerable impact than the square wave.

The extra time-to-stop of the triangular wave also gives a slightly reduced speed of second collision for unrestrained occupants so that, from these general considerations, the triangular wave appears to be better than the square wave. However, it must be borne in mind that real accidents occur at a range of speeds and the possibility cannot be excluded that future legislation may require tests at different speeds.

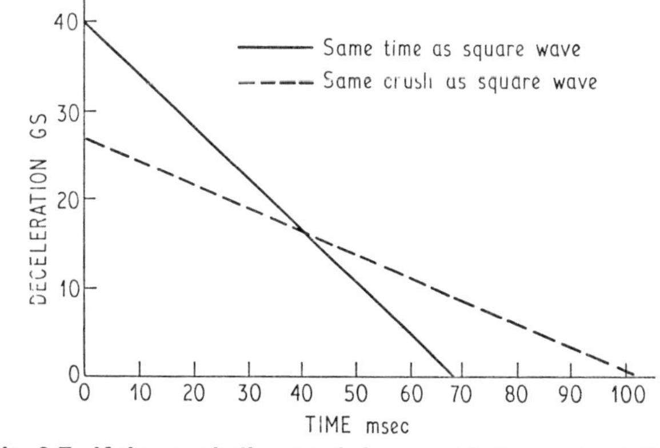

Fig. 2.7. If the crush distance is increased, the straight-line deceleration characteristic has a slower Severity Index than the square wave in Fig. 2.6, because impact time increases

It seems desirable that decelerations in low-speed impacts should be less severe than in high-speed impacts, but the concept of both square and triangular waves is based on an instantaneous or very rapid onset of deceleration which is likely to make low-speed impacts almost as severe as high-speed ones. If the triangular pulse is designed for a particular speed, impacts at higher speeds will result in the passenger compartment having to deform to absorb the excess kinetic energy, possibly resulting in loss of windshield, door opening, etc.

A suggested pulse for a 30 mile/h impact is shown in Fig. 2.8. This has a Severity Index of 113 and the total crush distance is 19·37 in, which is comparable to the 30 mile/h crush shown by tests on production vehicles.

Forces

The barrier impact simulation programme also calculates the forces from the deforming elements transmitted to the body structure. As these forces are known at particular instants, and the decelerations of the non-deforming parts of the vehicle are known for the same instants, the loads and stresses in the beam elements comprising the non-deforming 'passenger box' can be estimated, assuming that the structure remains elastic.

The calculations can be carried out by a programme designed to predict loads and stresses resulting from an external load system. Many such programmes of various degrees of sophistication have been written (**2**). Our own programme uses only idealized beam elements which may have offset ends (rigid gussets), and flexible connections between beams and nodes. Up to 43 nodes can be included without having to use external computer storage. Displacement restraints can be placed on any combination of nodes.

To use this programme, the 'passenger box' must be idealized into point masses, acting at the nodes. These masses produce external inertia forces acting at the nodes appropriate to the deceleration being considered. The remaining external forces are those produced by the deforming elements in the front of the vehicle. Enough restraints must be imposed on the structure to prevent it spinning round of flying away.

It is hoped that these programmes will improve understanding of the mechanics of barrier collision and provide some explanation of the sequence of events shown on films of impact, as well as giving guidance in the design of new vehicles.

Fig. 2.8 shows that, if maximum deceleration in the modified triangular pulse occurs at 0·010 s, the corresponding crush distance is only 5 in. In the simulation of the *Viva*, the driveline element does not exert a decelerating force until 0·025 s have elapsed and the dash-panel is not touched until 0·031 s after impact. It is clear, therefore, that the energy-absorbing part of the body structure must be carried right up to the front of the vehicle, so that the side rails, front fenders and wheelhouses can start decelerating the body effectively before valuable crush-distance has been used up.

The simple models described here can, of course, be made more complex. Sophisticated elastic load and stress-analysis programmes, incorporating tapered beams and membrane shear elements, have been developed, but they do involve considerable effort in collecting data.

The dynamic model could also be elaborated. The occupants and the rear unsprung mass could be included separately, for instance, and the front-axle mass broken down into a central mass and wheel masses: these have quite a large displacement relative to each other, as can be seen in films of impacts.

On some vehicles, the rotation of the engine–gearbox unit may be important, and this could be included in a more complicated dynamic model. There are potential difficulties here, as the decelerating force of the driveline would depend on both the rearward displacement and rotation of the engine–gearbox unit, relative to the body. These two variables would also have an effect on the impact of the engine unit on the dash panel.

There is clearly plenty of scope for different dynamic models, and it is hoped that, as more work is done on them, the more important features and modes of motion will become clear.

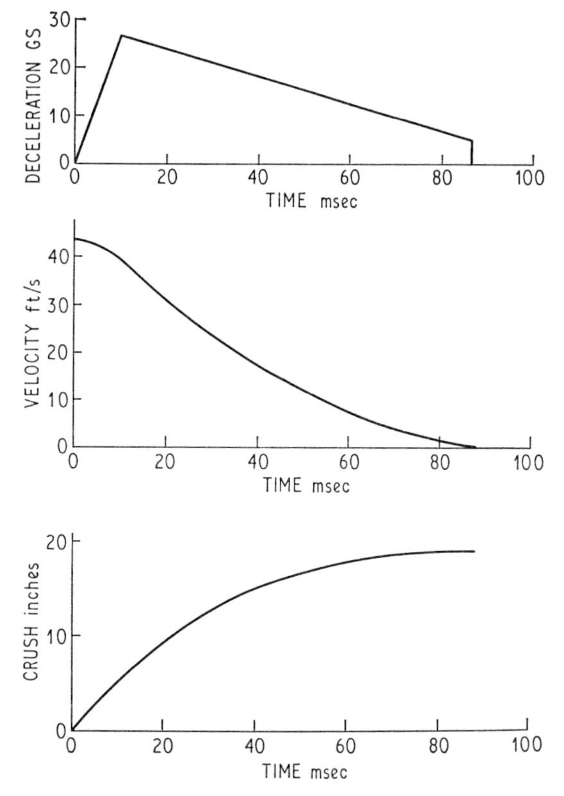

Fig. 2.8. Suggested optimum characteristics for deceleration, velocity and crush in a 30 mile/h barrier impact

MOVING BARRIER REAR IMPACT

The objective of this test is to avoid rupture of the fuel tank. Therefore it is obvious that, if the tank is placed well forward of the impact area (for instance, under the rear seat, or between the rear wheels), the chances of deformation and rupture are considerably reduced. This requirement of fuel tank protection points to the rear-end crush being minimized.

Reduced rear-end crush, however, will naturally lead to increased accelerations of the vehicle during the collision, with the consequent possibility of occupant injury. These conflicting requirements need to be optimized, which will be easier if the available crush distance is as large as possible. This can be achieved either by moving the fuel tank forward, or extending the rear-end.

The rear of the vehicle does not break down as conveniently as the front into elements for building up a mathematical model. One method would be to crush the complete rear-end with a static crusher and feed the resulting force/crush curve into a simple simulation programme, using only two masses. This sounds like almost as much work as actually carrying out a real impact test, but it would offer the advantages of strict control of impact speed, and the possibility of simulating impacts at various barrier speeds.

This model is shown in Fig. 2.9 and its equations of motion are:

$$M_1\ddot{x}_1 = -F$$
$$M_2\ddot{x}_2 = F - M_2 g$$

with initial conditions

$$x_1 = x_2 = 0$$
$$\dot{x}_1 = V$$
$$\dot{x}_2 = 0$$

again, to prevent rebound, $\ddot{x}_1$ may be set to zero when $\dot{x}_1$ reaches zero.

SIDE-IMPACT

Data collected by the M.O.T. Area Road Safety Boards indicates that, whilst side-impact accidents are quite infrequent, the percentage of serious injury is high. There seems to be a strong possibility that legislation will eventually set some standard of minimum deformation to the side of the vehicle when it is subjected to a standard test, either dynamic or static.

Neilson (1) has devised a simple mathematical model of side-impact which indicates the speed with which the occupants strike the interior for various impact speeds and initial portions. However, in severe impacts collapse of the door implies the possibility of the occupant being trapped in the car by the distortion of the door and side structure.

The problem presented to the body engineer is thus quite different from that of front or rear impact. There, the considerable depth of the structure can absorb energy in a predetermined way to protect the restrained vehicle occupants. By contrast, the side impact depth is so limited that there is little possibility of using the structure for energy absorption. In fact, the major requirement for the side structure is that it should deform as little as possible, thus deflecting the impact vehicle.

One way this can be done is by strengthening the doors and body side structure. The doors themselves must be reinforced, and care taken in designing the hinges and latches to ensure that these are not the weakest parts of the structure.

It is very difficult to determine in advance what the performance of a vehicle will be when subjected to side impact. At present, the method is to carefully examine tested vehicles to discover the mode and sequence of failure, and make appropriate modifications. A designer working on a new vehicle must, therefore, rely to a considerable extent on experience gained from tests on existing cars, and bear in mind that modifications and reinforcements may be necessary to the new design if test results are not up to expectations.

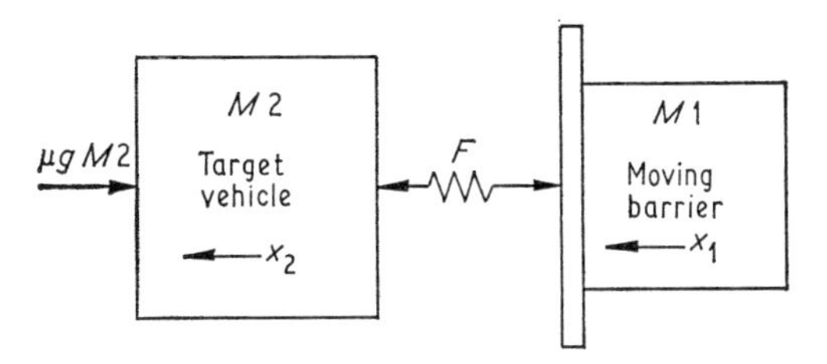

Fig. 2.9. A simulation model for a moving-barrier collision at the rear-end makes it possible to predict the effects of speeds

CONCLUSION

The performance of the vehicle in the barrier frontal impact test is at present the most important of the requirements of safety legislation as far as body structures are concerned, and is likely to remain so in future.

General Motors Research have shown that simulation of barrier impact by a combination of mathematic model and empirical data can give a guide to the performance of an existing or projected vehicle.

It is hoped that this technique will be further refined in future, perhaps with more complex idealizations, including, for instance, the vehicle occupants. With the increasing availability of accident data and the attendant growth in safety legislation, the body designer is likely to be faced with new challenges in the area of the side impact, rear impact, and increased speed of frontal impact. Finding solutions to these problems will mean that safety considerations will become even more fundamental in vehicle body design.

APPENDIX 2.1

REFERENCES

(1) NEILSON, I. D. 'Simple representations of car and unrestrained occupant impacts in road accidents'. Road Research Laboratory Report LR 249.

(2) ALLWOOD, R. J. and NORVILLE, C. C. 'The analysis by computer of a motor car underbody structure'. *Proc. Inst. Mech. Engrs* 1965–66.

(3) KAMAL, M. M. 'Analysis and simulation of vehicle to barrier impact'. SAE Report No. 700414.

(4) GADD, C. W. 'Use of a Weighted-Impulse Criterion for Estimating Injury Hazard'. Proc. 10th Stapp Car Crash Conference 1966.

Paper 3

A STUDY OF IMPACT BEHAVIOUR THROUGH THE USE OF GEOMETRICALLY SIMILAR MODELS

G. W. Barley*   B. Mills†

The behaviour of motor vehicles in frontal impacts has been investigated through the use of geometrically similar models. A large number of thin-walled cylinders have been impact-tested to establish scaling criteria and it was considered that models could be usefully employed in predicting full-size vehicle behaviour.

INTRODUCTION

THE SAFETY of the occupants of a vehicle during a collision depends partly on the 'deceleration signature' of the passenger compartment and partly on the crush characteristics of various elements of the structure. A recent paper by Grime and Jones (1)‡ deals with the movement of cars and their occupants after collision and makes the point that the front of the vehicle should have crushing characteristics more favourable to seat belt wearers and drivers. Work by the Digitek Corporation (2), Johnson and Wiltse (3) and also Kaufman and Larson (4) has stressed the importance of optimizing deceleration patterns.

Due to its complex nature a motor vehicle does not lend itself to ready analysis of impact behaviour but the study of simple structures, leading to scale model vehicle tests, does allow an appreciation of the nature of the problem. Several detailed analyses of impacts have been attempted in the past but it is suggested that principles to apply at the design stage have still not been established. G. C. Kao (5) describes an extensive investigation into 1/10th scale models and concludes that the reasonable results obtained, together with the obvious economies, make model tests worth while.

A number of mathematical models have been developed, one of the most recent ones being by Tani and Emori (6). Even simple mathematical models can provide a reasonable understanding of crush behaviour. A sophisticated approach is described by Thompson (7) in which a structural analysis programme, originally developed for structural elements working in the elastic region, has been extended to the plastic region to show behaviour during crushing.

Investigations have been made, by Postlethwaite and Mills (8), into the static and dynamic behaviour of simple structures during crushing. These included rods, and various sheet metal structures in the form of cylinders, cones and rectangular tubes, with and without discontinuities. The present paper describes an extension of this work to the investigation of the characteristics of a large number of cylinders of varying lengths, diameters and wall thicknesses to determine the effects of scaling and to assess the tolerance which should be applied to subsequent predictions. In addition, a linear accelerator has been designed and built which is capable of accelerating a 1000 lb mass to 44 ft/s, in order to test half-scale vehicle front ends. This represents a small scale barrier collision test facility. Consideration has been given to frontal impacts only. Model velocity and deceleration are measured and a high-speed film made of each test.

NOTATION

- A Cross-sectional area.
- E Young's modulus.
- h Wall thickness.
- I Second moment of area about neutral axis.
- L Length of impacted structure.
- M Mass of impacting body.
- n Number of half waves ($= 2L/\lambda$).
- r Radius of tube ($= D/2$).
- V_0 Initial impact velocity.
- V_1 Velocity at yield point.
- V Instantaneous velocity.

The MS. of this paper was received at the Institution on 28th May 1970 and accepted for publication on 28th May 1970.

* *Body Engineer, Rootes Motors, Coventry.*
† *Senior Lecturer, Department of Mechanical Engineering, University of Birmingham.*

‡ *References are given in Appendix 3.1.*

Y Yield stress.
λ Convolution wave length.
ρ Density.
σ Axial compressive stress.
σ_c Critical value of σ from classical theory.
ω Natural frequency $(=\sqrt{AE/ML})$.
$\bar{y}_0$ Initial lateral displacement for $x = \lambda/4$.
$\bar{y}(t)$ Lateral displacement at time t for $x = \lambda/4$.

THEORY

For the purpose of analysing a vehicle body-structure, the behaviour of a number of basic structures needs to be known. A front-end car structure generally includes rectangular sidemembers and open sections formed from the wings and valances. The former may be analysed directly and the latter considered equivalent to rectangular tubes with large cut-outs. Since the behaviour of the latter can be derived from that of thin-walled cylinders, equations are presented here for the analysis of these simple structures; the derivation of the equations may be found in the paper by Postlethwaite and Mills (**8**).

Cylinder

This is analysed in two parts, the pre-failure phase and the post-failure phase.

Pre-failure phase

The cylinder is assumed to behave such that the form of failure is sinusoidal (Fig. 3.1)

$$y = \frac{\bar{y}(t)}{n}\cdot\sin\left(\frac{n\pi x}{L}\right)$$

The general equation for lateral motion of any element is

$$\frac{d^2\bar{y}(t)}{dt^2} - (B_1 \sin\omega t - B_2)\bar{y}(t) = B_3 n\bar{y}_0$$

where

$$B_1 = \frac{n^2\pi^2}{A\rho L^2}\cdot MV_0\omega$$

$$B_2 = \left(P_E - \frac{2\pi EhL^2}{rn^2\pi^2}\right)\frac{n^2\pi^2}{A\rho L^2}$$

$$B_3 = \frac{n^2\pi^2}{A\rho L^2}\cdot P_E$$

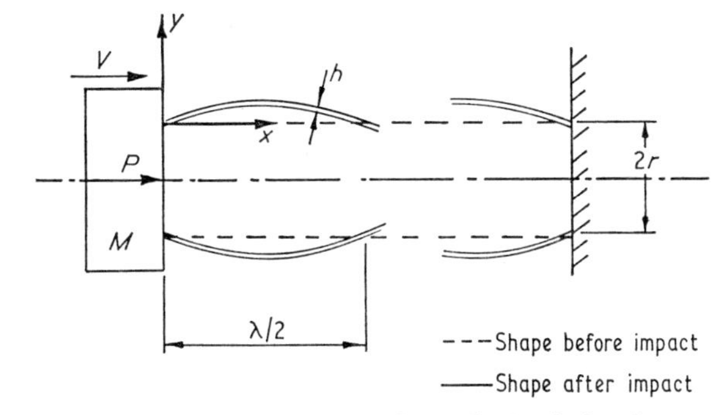

Fig. 3.1. Pre-failure phase of a cylinder being impacted

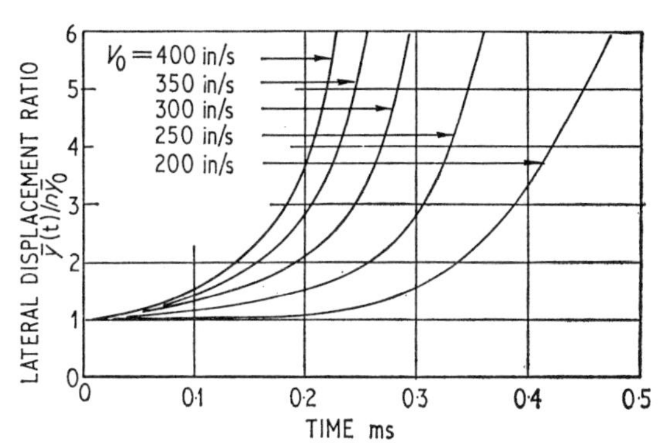

Fig. 3.2. Lateral displacement ratio against time at various impact speeds as calculated from the general equations

where

$$P_E = \frac{n^2\pi^2}{EI/L^2}$$

The equation of motion can be solved on a digital computer for various values of V_0, a typical family of curves being shown in Fig. 3.2.

Failure of the cylinder would occur at $\bar{y}(t)/n\bar{y}_0 = \infty$ if classical theory was applied, but due to imperfections in manufacture σ/σ_c is taken as 0·66 leading to $\bar{y}(t)/n\bar{y}_0 = 3$ (**9**). The time at which $\bar{y}(t)/n\bar{y}_0 = 3$ can then be substituted in $V = V_0 \cos\omega t$ giving the trolley velocity at which failure occurs.

Post-failure phase

A cylinder will collapse in the form shown in Fig. 3.3. The idealized form shown has been analysed (**10**) and resulted in an expression for the mean load during crumpling of

$$\bar{P} = KYh^{1\cdot5}\sqrt{2r} \qquad \text{where } K \simeq 6$$

This expression, evolved for static collapse of a cylinder, has been found to be applicable to the post-failure phase of a dynamic test. The post-failure deceleration is given by $\ddot{x} = \bar{P}/M$, and the energy absorbed by $\bar{P}.x$.

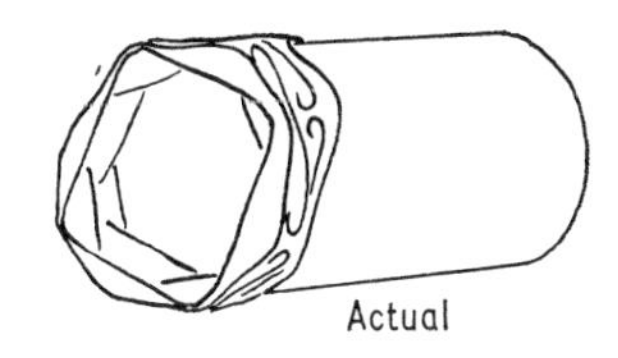

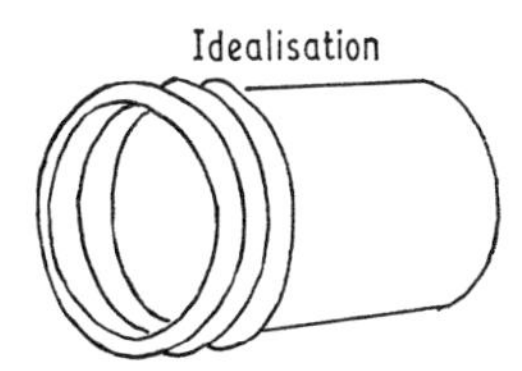

Fig. 3.3. How a cylinder actually collapses, as against the idealized crushing shape assumed in the analysis

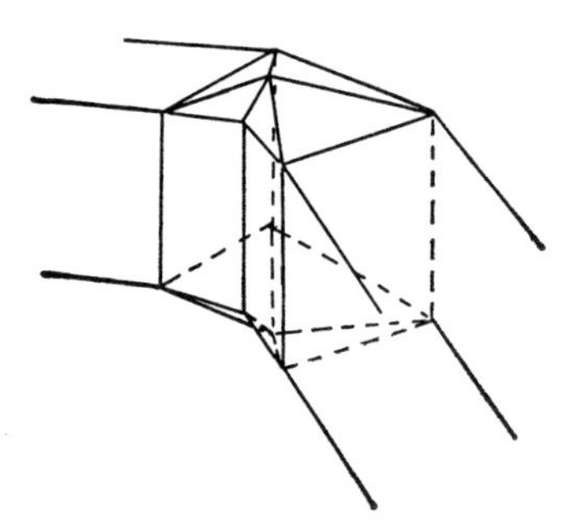

Fig. 3.4. The energy absorbed by a collapsing rectangular strut can be predicted from the angle through which the various instantaneous 'hinges' turn at the 'elbow'

Cylinders with discontinuities

It has been shown (**8**) that cylinders with cut-outs behave in the same way as those without, provided the cut-out does not exceed 50 per cent of the cylinder circumference.

Rectangular tubes

Previous work (**8**) has shown that these can be considered as behaving in the same way as cylinders of equivalent perimeter, except where deformations greater than 20 per cent of the length occur.

Struts of rectangular section

A geometrical analysis is being carried out which predicts the energy absorbed in a strut by considering the angle through which the various hinges turn at the elbow (see Fig. 3.4). This type of failure occurs in structures with the proportions of vehicle sidemembers.

Scaling

The energy involved in the elastic deformation of a sheet-metal structure is often small compared to that for the subsequent plastic deformation. Some idea of the scale effect can therefore be obtained by considering only the plastic deformation in the post-failure phase. Further justification for this approach is that many elements of a vehicle structure are not ideal sections but involve some pre-buckling so that the elastic phase does not exist.

Considering a cylinder, the mean load during crumpling in the post-failure phase is quoted as $P = KYh^{1\cdot5}\sqrt{2r}$ and the energy absorbed as $P.x = KYh^{1\cdot5}\sqrt{2r}.x = MV^2/2$. Considering further two cylinders suffixes 1 and 2,

$$\left(\frac{h^{1\cdot5}\sqrt{r}.x}{MV^2}\right)_1 = \left(\frac{h^{1\cdot5}\sqrt{r}.x}{MV^2}\right)_2$$

If the velocity effect is ignored then this is satisfied by making,

$$h_1:r_1:x_1:L_1 \equiv h_2:r_2:x_2:L_2$$

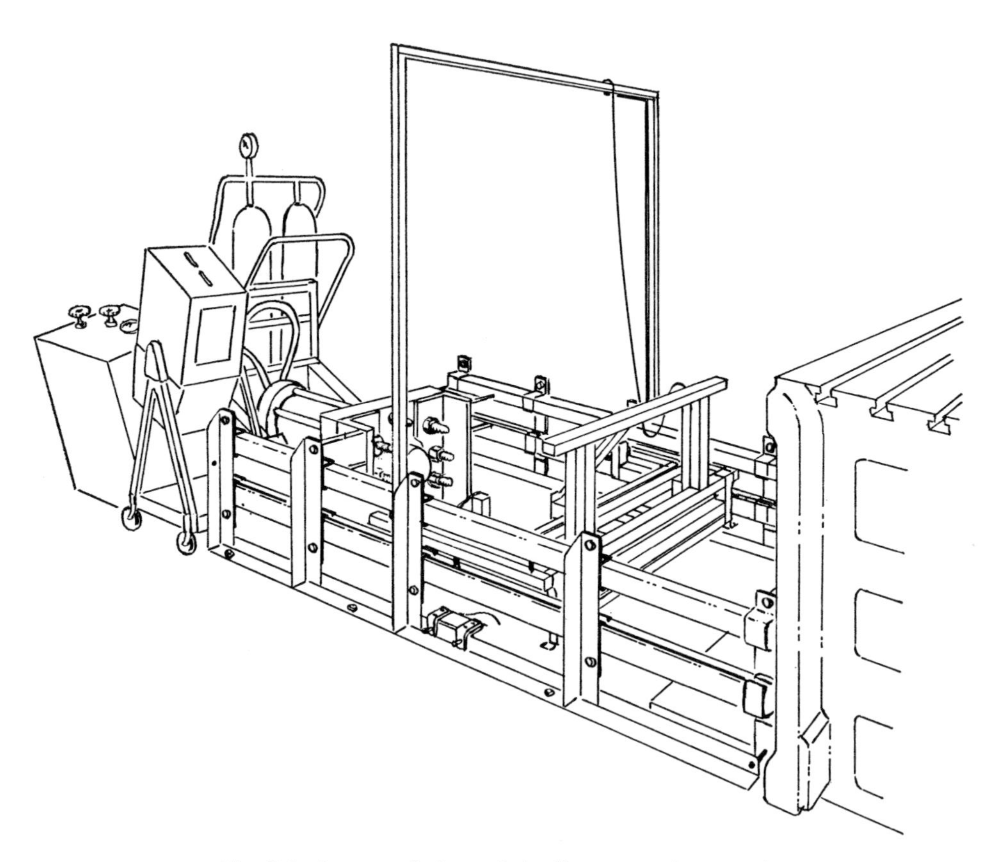

Fig. 3.5. A general view of the linear accelerator rig

and
$$\frac{M_1}{L_1^3} = \frac{M_2}{L_2^3}$$

A half-scale model (all linear dimensions, including h, halved) would exhibit the same percentage reduction in length as full scale when impact is at the same velocity but with one-eighth of the mass.

DESIGN OF LINEAR ACCELERATOR

A general view of the rig, cross-section of the air-ram and details of the air-circuit appear in Figs 3.5, 3.6 and 3.7 respectively.

The air-ram was designed to work on the *Dynapak* principle with a maximum working pressure of 1500 lbf/in². Air was considered to expand adiabatically within the cylinder; an energy balance appears below,

$$\underset{\text{available}}{E} = \underset{\text{required}}{E_1} + \underset{\text{dissipated in ram}}{E_2}$$

which can be written in the numerical form below, using the chosen dimensions of

$$a = 10{\cdot}5; \quad b = 8{\cdot}25; \quad d = 5{\cdot}197$$

$$312{\cdot}4(65{\cdot}4\,.\,S+82{\cdot}8)\left[1-\left(\frac{65{\cdot}4\,.\,S+82{\cdot}8}{86{\cdot}7\,.\,S+82{\cdot}8}\right)^{0{\cdot}4}\right] = 30\,080+(29{\cdot}6\,.\,S+1637)$$

where S is the stroke of the piston.

Consideration of the following factors led to the choice of cylinder dimensions used:

1. Availability of materials
2. Cost
3. Acceleration of ram to be minimized

} Favour $c{:}d$ large

4. Space available
5. Rigidity of ram tube
6. Sealing of ram head against main cylinder end cap

} Favour $c{:}d$ small

7. Dimensions of proprietary articles (e.g. ring springs)

Requirements 3 and 6 led to the choice of two bearings to support the ram, spaced about 9·5 in apart and separated by two concentric rows of ring springs. These fulfil the important function of dissipating the energy of the ram head and ram tube at the end of its stroke. One bearing floats and moves a maximum of 2 in compressing the ring springs. These are steel rings with tapered faces which alternately expand and contract (radially), absorbing some $\frac{2}{3}$ of the input energy; they limit the maximum trolley speed to 44 ft/s.

A few problems arose in the manufacture of the air ram, the chief one concerning the tube which had been designed to 5·197 in dia., 0·25 in wall thickness, in the interests of

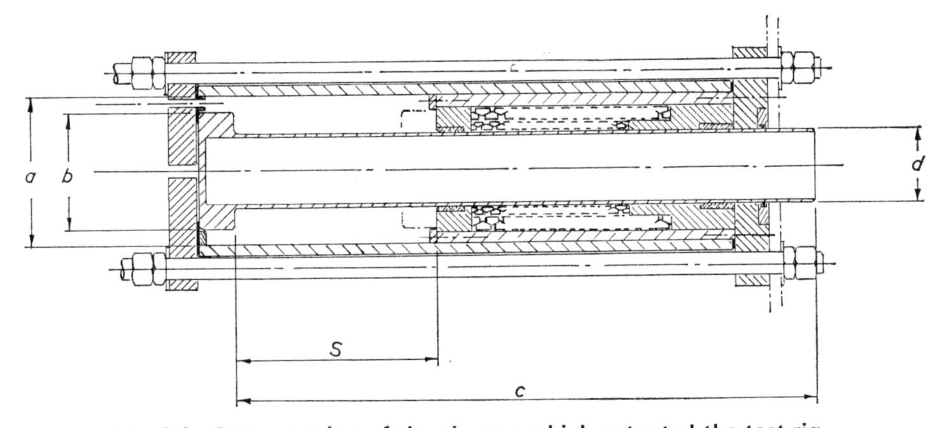

Fig. 3.6. Cross-section of the air-ram which actuated the test rig

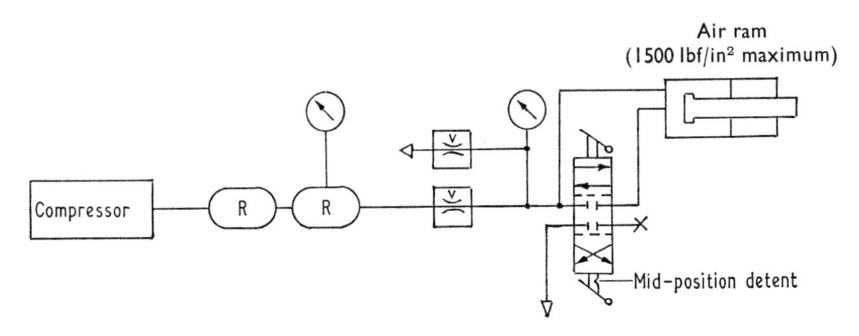

Fig. 3.7. The air-circuit which powers the accelerator rig

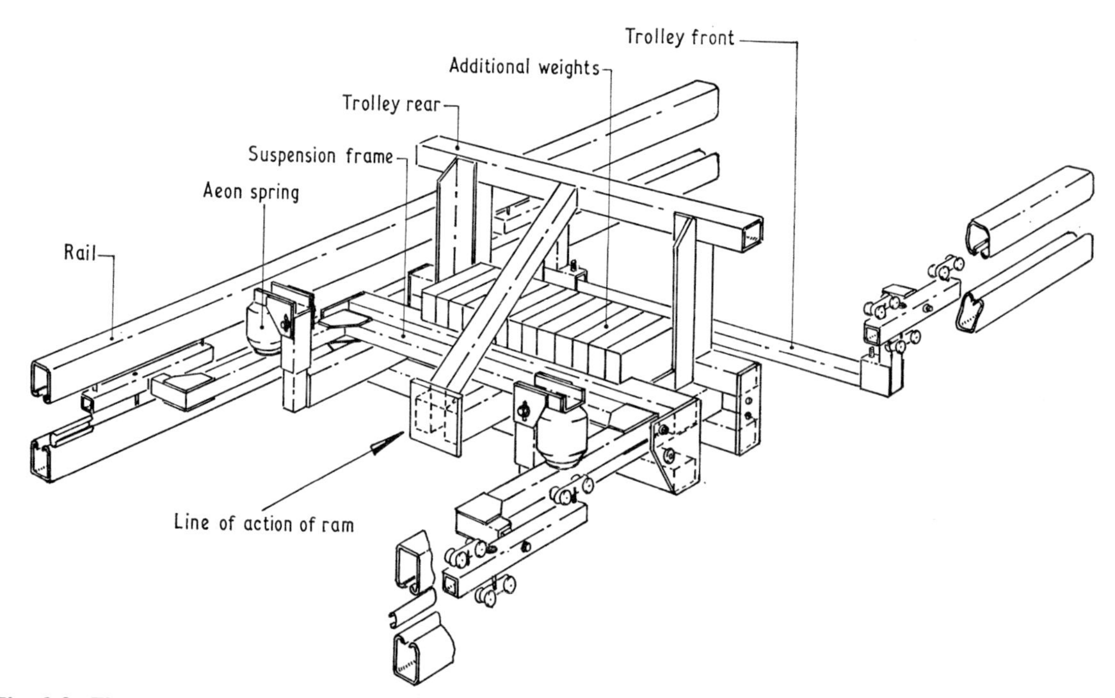

Fig. 3.8. The trolley system of the accelerator was designed for mounting the model shown in Fig. 3.11

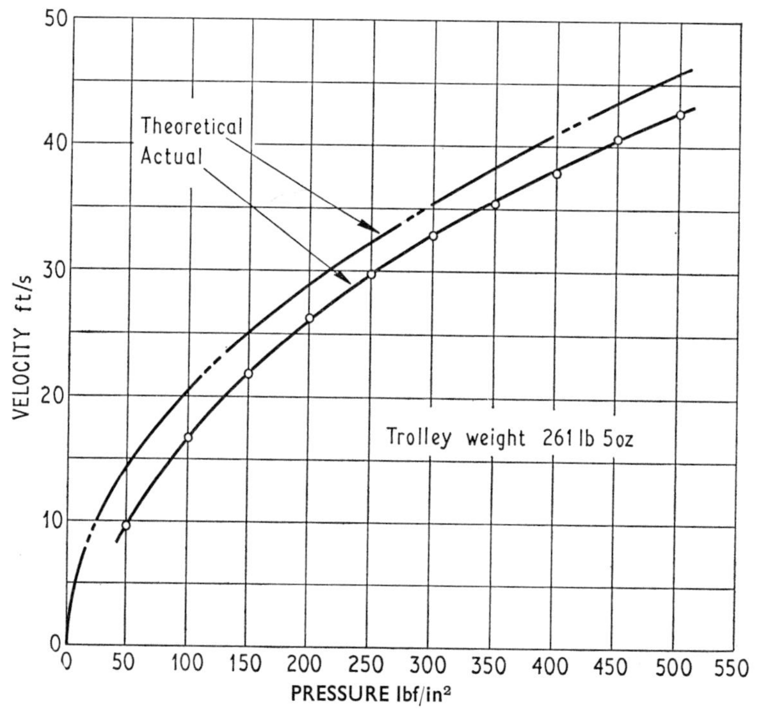

Fig. 3.9. Performance characteristic of the linear accelerator

reducing mass but was far from ideal for obtaining the correct bearing clearance because some ovality seemed unavoidable. Considerable care was also needed in making the bearings colinear.

The trolley (Fig. 3.8) was designed specifically for mounting the model shown in Fig. 3.11. The latter was supported at the front suspension point on the front trolley, the rear trolley being attached by means of a plate bolted to the sills, sidemember extension and gearbox cover and by two plates attached to the upper 'A' posts. It was anticipated that, with this method of mounting, the load paths in the model would faithfully represent those in the full-size vehicle.

A further innovation enabled the model to tilt forwards on contact with the block so as to simulate this frequent occurrence in full-scale vehicle impacts. It will be seen that the model, front trolley and rear trolley may be considered as one body, pivoted at the front trolley mounting points. The rear trolley is supported on *Aeon* springs, bearing on the suspension frame. The model can tilt freely through 27° before the suspension frame limits further upward movement of the rear trolley. This construction allows the wheels on which the trolley moves always to remain in contact with the rails but gives the desired freedom of movement.

The linear accelerator was calibrated in terms of pressure versus velocity for a given trolley mass, a typical curve appearing in Fig. 3.9. Velocity was obtained from the oscilloscope display of the output of a photocell fixed to the side of the rig, and excited by a black and white chequered plate fitted to the trolley. It was found that trolley velocity could be accurately specified in terms of cylinder pressure.

EXPERIMENTAL METHOD

Cylinder tests

Static tests of cylinders were carried out on an *Instron* machine. Load versus deflection curves were obtained which allowed comparison of the mean load after yield with Alexander's prediction. Tests were also carried out on steel specimens made to BS 18 to determine the value of the yield stress Y.

Dynamic tests of cylinders were carried out on a small linear accelerator (**8**). The air-ram was calibrated in terms of pressure versus trolley velocity for various values of trolley mass, so that the required combination of mass and velocity could be readily obtained.

A UAI strain gauge accelerometer with an undamped natural frequency of 800 Hz was attached to the trolley, the output being fed into an oscilloscope, and a photograph was taken of the display. A 300 Hz filter was generally used to remove the signals due to stress-waves traversing the trolley.

Measurement of the total crush was taken to represent the post-failure phase only, as little deformation takes place in the pre-failure phase.

Scale model tests

These were carried out in a manner similar to the dynamic tests on cylinders but with more elaborate instrumentation. Trolley velocity was monitored on a second oscilloscope and a Fastax camera film record taken at 2000 frames/s. (The camera was manually started approximately 0·5 s before the ram was fired.)

EXPERIMENTAL RESULTS AND DISCUSSION

Cylinder tests

The scope of cylinder tests is shown in Table 3.1

Table 3.1. Properties of cylinders tested

	Material thickness, in 0·014, 0·222, 0·028, 0·035 and 0·049				
D, in . .	1.43	2·25	2·86	3·57	5·00
L, in . .	2·86 and 4·28	4·49 and 6·73	5·72 and 8·57	7·14 and 10·72	10 and 15

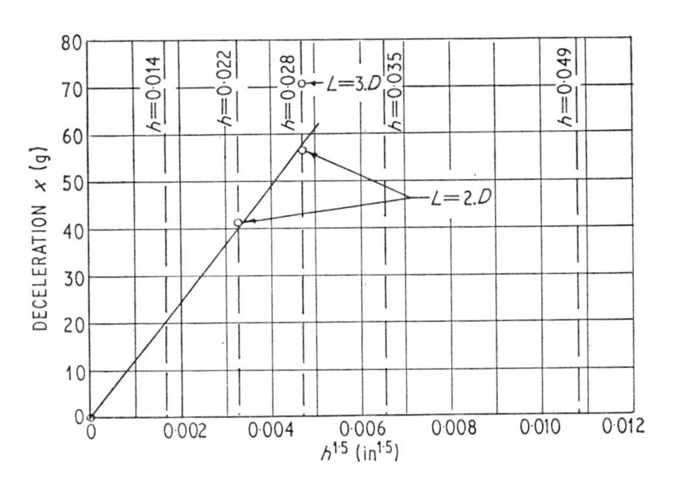

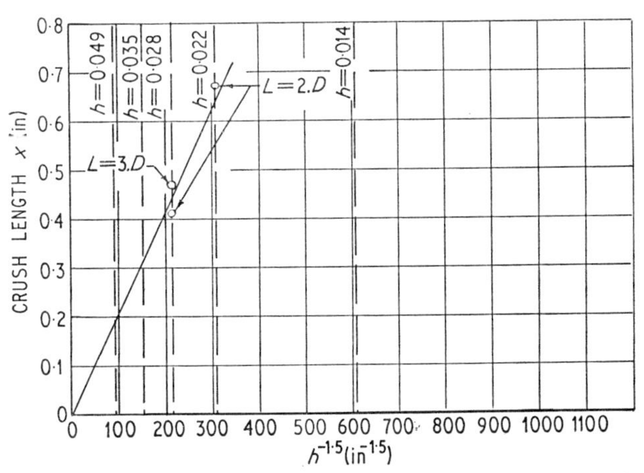

Fig. 3.10. The approximation to straight lines of the cylinder test results shows that it is reasonable to ignore the pre-failure phase of crushing

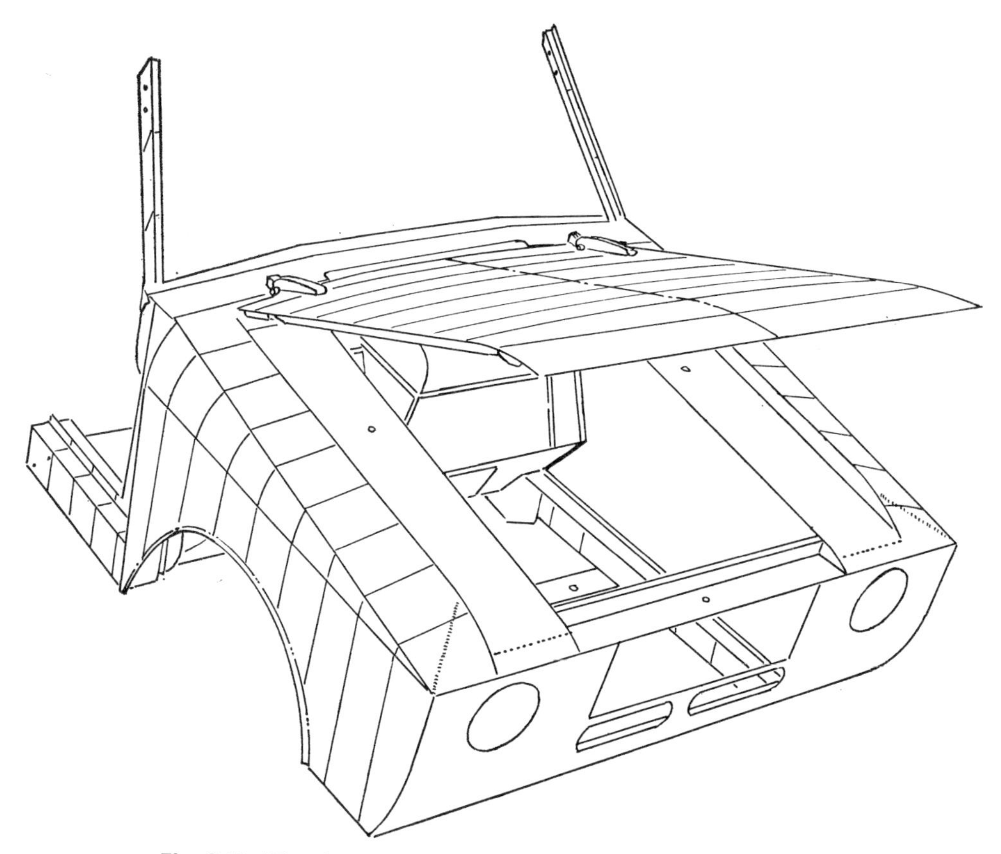

Fig. 3.11. The shape of the half-scale model which was tested

It is worth considering that, if the pre-failure phase is completely ignored and the form of the post-failure phase taken to describe the whole crush behaviour, then the following would apply:

(i) if $M \propto D^3$, V constant, $h \propto D$ — then $\ddot{x} \propto D^2$ and $x \propto D$

(ii) if $M \propto D^3$, V constant — then $\ddot{x} \propto h^{1\cdot5}$, $x \propto I/h^{1\cdot5}$

Plotting the quantities on the r.h.s. would lead to a straight-line graph in each case, if the assumption is reasonable. Fig. 3.10 shows some results obtained, which suggest that as a first approximation this approach may be made.

Half-scale model tests

The model (Figs 3.11 and 3.12) was based on a conventional saloon but with the structure simplified and styling considerably altered by the use of only single-curvature panels.

The steel used in construction was of identical composition to that used for vehicle bodies but the panel thickness was approximately half that of the corresponding full-size panel.

Models were assembled by using a hand spot welder, the intention being that failure would always be by distortion of a panel rather than by breaking open of a spot weld.

The results of two tests are currently being analysed. They were performed under the following conditions:

(1) No allowance for model tilting and representation of vehicle bodywork only—i.e. no engine simulation etc., to simulate correctly the load-paths and mass distribution.

(2) As (1), but front trolley and rear trolley joined together, model supported only on rear trolley.

In both of these tests the energy input to the trolley has been approximately one-eighth that of the full-size vehicle in a normal barrier impact.

CONCLUSIONS

A rigorous analysis of vehicle-impact behaviour is a very difficult task. The use of models and testing of simple structures can provide a basis for predicting the performance of complex structures which will be useful in spite of the unrefined approach adopted in the work described here.

Reduction of scale and energy requirements effects considerable economies and allows a comprehensive investigation of the problem. The knowledge gained from the investigation should lead to some useful guidelines

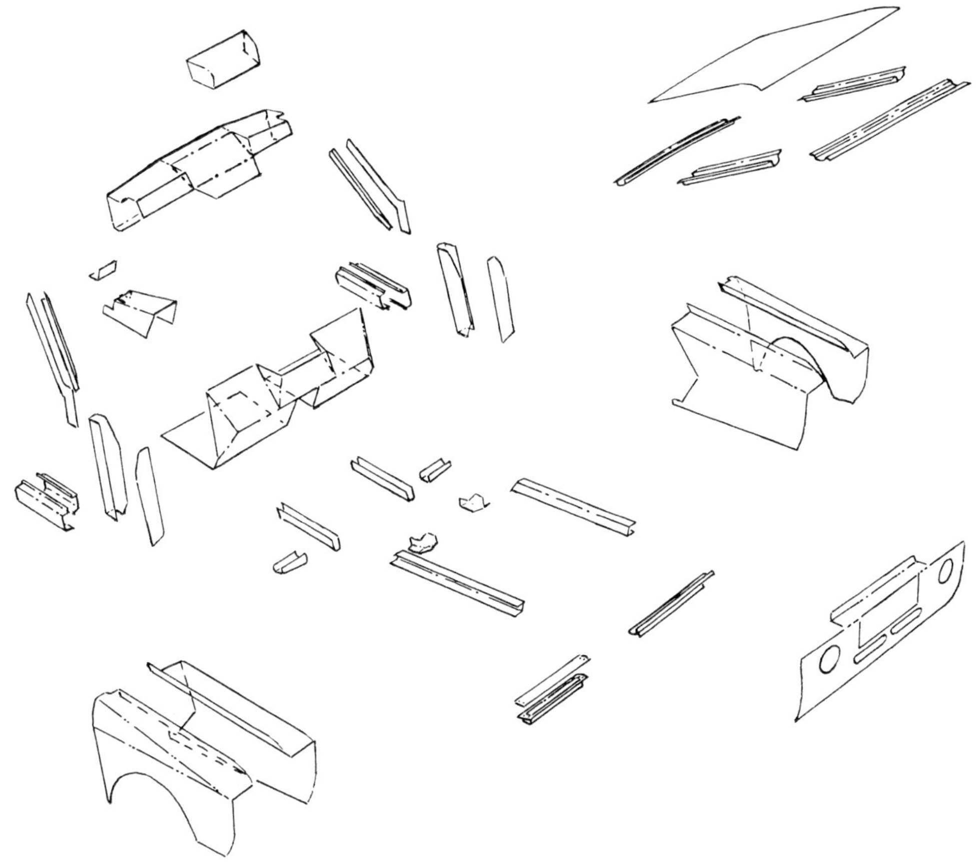

Fig. 3.12. The parts of which the half-scale model front-end consisted

which may be applied at the initial body design stage and suitable optimization of design provide greater protection for occupants in the event of an accident.

ACKNOWLEDGEMENTS

The authors wish to thank Professor S. A. Tobias for the provision of laboratory facilities, and Rootes Motors Limited for sponsorship of the project.

APPENDIX 3.1

REFERENCES

(1) GRIME, G. and JONES, I. S. 'Car collisions—the movement of cars and their occupants in accidents', I.Mech.E. AD P5/70.

(2) DIGITEK CORPORATION. 'Experimental safety vehicle program'.

(3) JOHNSON, P. R. and WILTSE, W. E. 'Front structural strength as it affects occupant injury reduction and survival'.

(4) KAUFMAN, H. and LARSON, D. B. 'Calculation of Deceleration Waveforms using optimal control theory'.

(5) KAO G. C. *et al.* 'A scale model study of crash energy dissipating vehicle structures'.

(6) TANI, M. and EMORI, R. I. 'A Study on Automobile Crashworthiness', SAE 700175.

(7) THOMPSON, J. E. 'Occupant Response versus vehicle crush: A Total System approach', SAE 680780.

(8) POSTLETHWAITE, H. E. and MILLS, B. 'Use of collapsible structural elements impact isolators, with special reference to automotive applications', *J. of Strain Analysis* Vol. 5, No. 1, 1970.

(9) DONNELL, L. H. and WAN, C. C. 'Effect of imperfections on buckling of thin cylinders and columns under axial compression', *Trans. Am. Soc. Mech. Engrs* 1950 **72,** 73.

(10) ALEXANDER J. M. 'An approximate analysis of the collapse of thin cylindrical shells under axial loading', *Q. Jl Mech. appl. Math.* 1960 **13,** 10.

Paper 4

TOWARDS THE ALL-PLASTICS MOTOR CAR

G. O. Gurney*

This paper outlines the wide spectrum of plastics now available and attempts to give a basic understanding of the techniques for this manufacture. Automobile engineers often say that the plastics industry does not appreciate the real problems of the motor industry. The author questions this and suggests that the motor industry does not understand the real advantages of sophisticated plastics.

WHY PLASTICS?

IT IS TRUE that the motor industry designed and built very successful automobile bodies before the development of plastics. Leather, cotton fabrics, horsehair and wool padding, and metals, were used for body construction long before plastics came along. There has, however, been a phenomenal growth in the use of plastics in automobile body manufacture in the last ten years.

The reasons are quite simple. Plastics have achieved:

(i) substantial manufacturing cost savings,
(ii) superior component performance at no extra cost,
(iii) weight savings, and hence improved total vehicle performance, at no extra cost,
(iv) a specific performance when no other material would suffice.

PRESENT SITUATION

Today's automobile designers are familiar with the broad range of components that can be manufactured in plastics, if not necessarily with the particular materials to be used. Table 4.1 shows the growth-rate of plastics in automobile manufacture and an interpretation of the minimum projected growth-rate to 1980, exclusive of body panel construction.

It can be seen that the recent growth rate has been substantial and, even if the future growth is restricted to components and body trim, plastics usage in car manufacture will about treble in the U.K. in the next ten years. However, that total tonnage is not very high, only about 5 per cent of the total national plastics consumption. A breakthrough into body-panel production, however, would quickly make the motor industry one of the largest users of plastics, probably on a par with the building industry.

One must then examine the reasons why manufacturers would wish to change from metal to plastics for body structures.

The world's supply of basic raw materials, far from being conserved, is being eaten into at a prodigious rate. The raw materials for steel are no exception. A number of calculations have been made as to when these materials are likely to become critically short in supply and generally they purport to show that, if the present rate of usage and wastage continues, steel production will be severely threatened before the year 2000 A.D. It is already recognized that motor-body manufacture is one of the most wasteful uses of steel.

The material wastage in plastics moulding is invariably below 2 per cent so that, from the point of material conservation alone, plastics become a major contender. Another point is economic viability. Even if it is assumed that new mineral resources will be found, allowing steel

Table 4.1. Past and projected growth rates

Year	U.K.			
	lb/car	m. cars /yr.	Total plastics, m lb/yr	
1955	4	0·9	3·6	
1960	12	1·3	15·6	
1965	30	1·7	51	
1967	40	1·5	60	
1968	48	1·8	81	
1970	55	1·9	105	Estimates
1980	85	3·5	300	Estimates

The MS. of this paper was received at the Institution on 24th June 1970 and accepted for publication on 11th February 1971.
* *General manager, Hills Precision Die Castings Ltd, Birmingham.*

production to continue unabated, the process of steel manufacture has probably approached its optimum economic level and the projected continuing price rise in steel must be considered in comparison with the plastics polymer prices which continually decrease as use increases. Therefore, the time is fast approaching when plastics panels will compete with steel on more favourable economic grounds. This time is drawing even closer due to the very substantial and exciting material and process developments being fostered by the ebullient and progressive plastics industry. Further savings in manufacture will accrue as the automobile designers become more familiar with the properties of the various plastics and also aware of the greater design freedom which can be achieved with plastic moulding.

COMPETING PROCESSES

In the U.K., plastic body panels are used for Lotus, Reliant, Marcos and similar cars as well as for a number of commercial-vehicle cabs. These are low production rate vehicles and the contact-moulded fibreglass construction used predominantly has perhaps little relevance for high production rates—say 2500+ vehicles per week.

Fibreglass bodies could be produced by the contact method at 2000+ per week but the vast labour force and floor area required, together with low reproducibility in large-scale production, would make this an unacceptable method for major motor manufacturers.

This leaves the following production processes, several of which are already commonplace in the plastics industry but which, in some instances, would have to be scaled up to cope with the larger size of panels required.

Injection moulding

In this process, thermoplastic materials, that is those materials which soften on heating, are fed from a hopper into a screw conveyer. Rotation of the screw moves the plastic granules or powders, just as in a giant mincing machine. As the materials are moved down the barrel, they pass through controlled-heating zones which ensure that by the time the material has been drawn down to the nozzle end, it is a melt of low viscosity.

At this stage, the screw, acting as a plunger, is pushed forward towards the nozzle, forcing the melt through the nozzle and into the cavities of a mould held together under hydraulic pressure.

The mould is refrigerated at the next stage, bringing the material below its softening point and, after a suitable cooling time has elapsed, the mould opens and the moulding is removed from the cavity.

Comparative disadvantages

(i) Heaviest investment in plant.
(ii) Heaviest tooling investment.
(iii) Some size limitations though normal auto-panel size presents little problem.

Advantages

(i) Fastest moulding cycles.
(ii) Wide choice of materials.

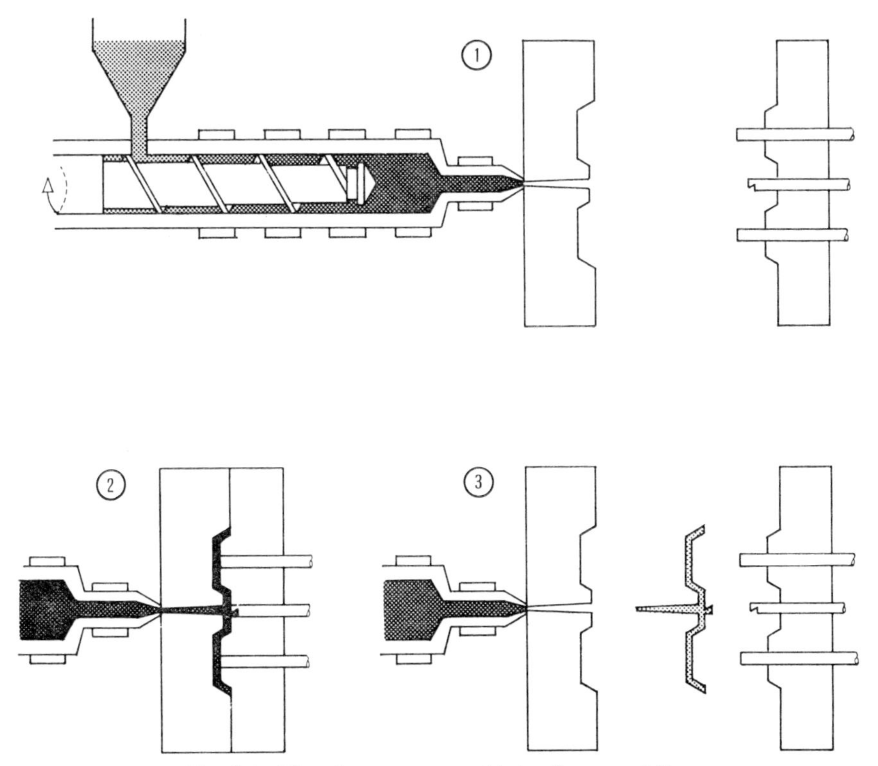

Fig. 4.1. The three stages of injection moulding

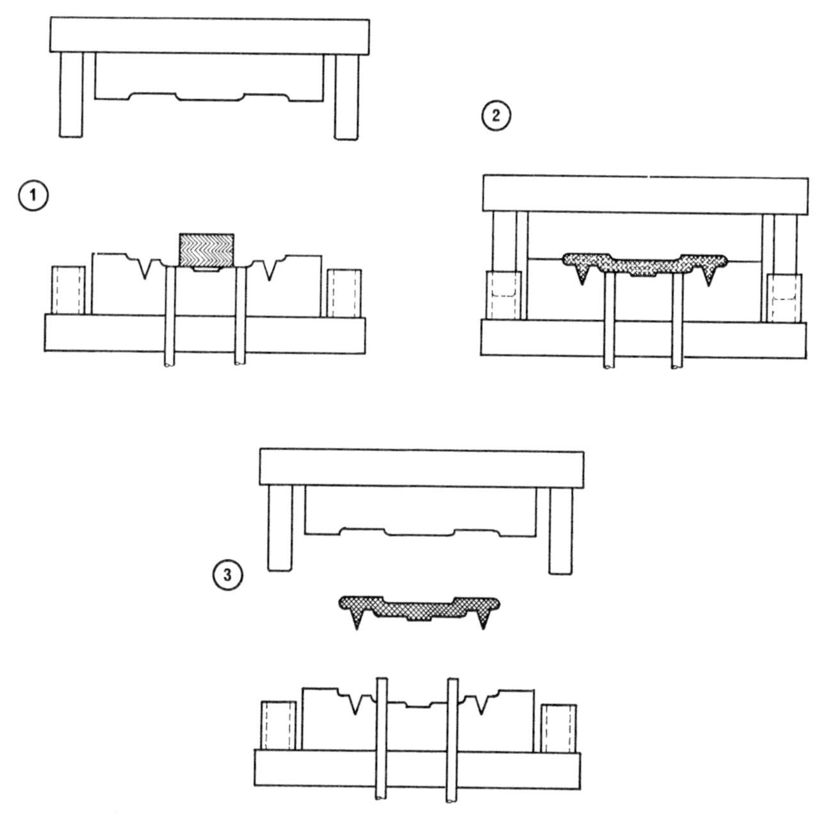

Fig. 4.2. In compression moulding a measured amount of material is placed between the open tools

(iii) Probably lowest unit-costs.
(iv) Inbuilt stiffening ribs and bosses.
(v) Excellent reproducibility.
(iv) Minimal factory area.

Compression moulding

For large mouldings manufactured in this manner, fibre-reinforcements are used—either glassfibre or sisal. In the latest techniques a measured weight of resin pre-impregnated fibrous material is placed between the open tools, the resin having previously been taken to a partially cured state so that it is tacky but not liquid. The metal moulds are pre-heated to around 150°C, and the hydraulic press then closes the mould, applying both heat and pressure to the material.

Initially the heat reduces the viscosity of the resin, enabling it to flow throughout the mould cavity under the application of the closing pressure. When flow is completed, the heat continues to act upon the resin which then hardens to form the moulding. These resins are called thermosetting as opposed to the thermoplastics which soften upon heating.

To complete the process when the hardening of the resin is completed, the moulds are opened and the moulding removed. This is the technique used in the manufacture of the Chevrolet *Corvette* car body.

Disadvantages

(i) Medium plant costs.
(ii) Medium tooling costs.
(iii) A limited number of materials can be processed.
(iv) Fairly high raw material cost.

Advantages

(i) Medium fast moulding cycles.
(ii) Built-in stiffening ribs and bosses.
(iii) Excellent reproducibility.
(iv) Minimal factory area.
(v) Fairly large panels can be moulded.

Blow moulding

The first part of the blow moulding process is virtually identical to injection moulding except that, instead of the thermoplastic material being forced through a nozzle into a mould, it is extended from the barrel through a die to form a tube.

These tubes (or parisons, as they are called) are fed between split moulds which close and compressed air is blown into the tube, forcing its walls outwards against the mould. Once the material has cooled, the mould opens and the hollow container is removed for further processing.

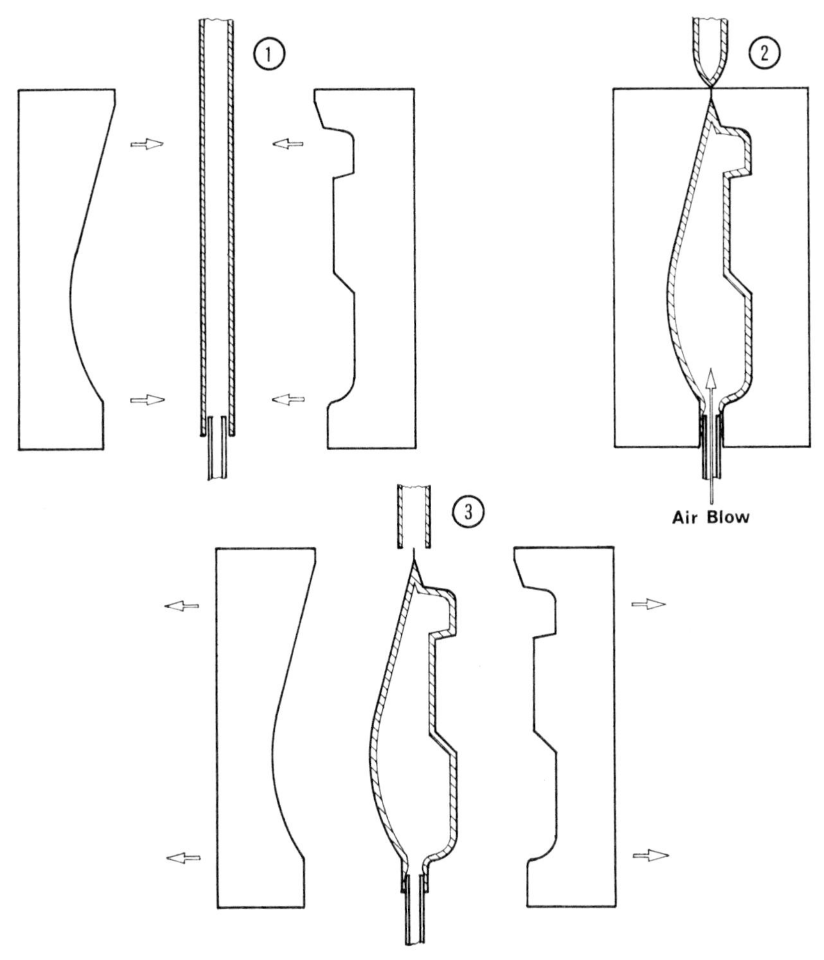

Fig. 4.3. Blow moulding has the advantage of cheap tooling and is suitable for double-skin panels

Disadvantages

(i) Heavy equipment costs.
(ii) Slower mould cycles than injection moulding.
(iii) Limited in size of application due to nature of process.
(iv) Difficulty in controlling thickness of panels.
(v) No ribs or bosses.
(vi) Strictly limited number of suitable materials.

Advantages

(i) Fairly cheap tooling.
(ii) Excellent for double-skin configurations such as doors, where inner and outer skins are formed in the same component.
(iii) Multiple panels could be blown and subsequently trimmed.

Rotational casting

There are several variations on this process but basically box-type moulds are rotated in all directions and a measured charge of plastics material inside the box spreads itself over the surface of the mould through centrifugal action. When the plastic is either cooled or set, according to its type, the mould is opened and the plastics item is removed.

Disadvantages

(i) Limited in application.
(ii) Longish moulding cycles, hence floor area and equipment requirements probably excessive for high production rates.
(iii) Probably highest process costs.
(iv) Difficulty in maintaining wall thickness and reproducibility.
(v) No stiffening ribs or bosses.
(vi) Limited range of materials.

Advantages

(i) Opens up possibility of large, one-piece body structures, albeit with limitations of design, thus giving greatly reduced body assembly costs.
(ii) Low tool investment.

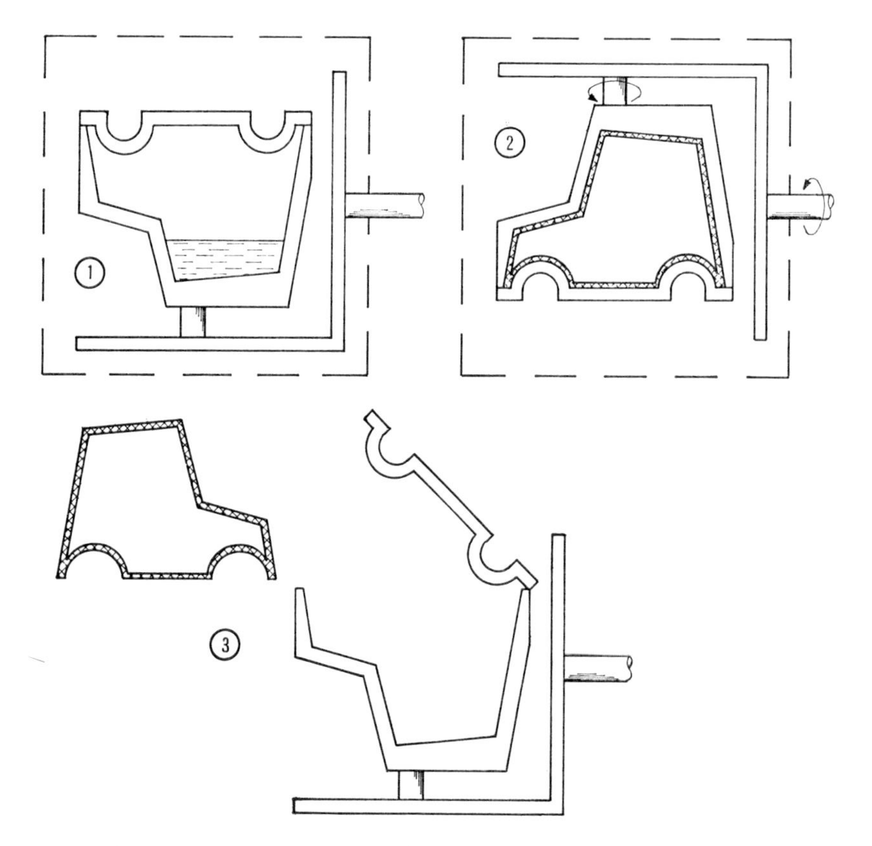

Fig. 4.4. Rotational casting is limited in application but permits one-piece fabrication of large pieces

Vacuum forming

This process is used to produce the body panels of that interesting vehicle, the Citroën *Mehari*. It was also used to produce the Cycolac research vehicles CRV 1 and 2 made by the Centaur Engineering Division of Marbon Chemicals Limited in Detroit.

Thermoplastic sheet materials are clamped round the periphery of a mould and heated to a flexible state. The sheet is then brought into contact with the mould and vacuum is applied to draw the sheet down tightly over the varying contours, after which a cooling cycle brings the temperature down to a point where the shape can be handled without deformation. After unclamping, the component is blown off the mould.

For large body panels, necessitating thick sheet, a 'plug-assist' device is used which presses the sheet into intimate contact with the mould, particularly in areas of deep draw.

Disadvantages

(i) No ribs or bosses.
(ii) The most expensive raw materials.
(iii) Limited range of materials.
(iv) Fairly long cycles.

Advantages

(i) Large panels possible, such as one-piece under-tray, etc.
(ii) Lowest tool-cost.
(iii) Low basic equipment-costs.

Injection casting

This is a relatively new process developed from the foam moulding technique used for the production of crash pads. Two reacting liquids are mixed in a predetermined ratio and a measured quantity is injected into simple closed moulds which are usually manufactured from glass-reinforced epoxide resin. The reaction between the two resins is exothermic and a cross-linked plastic is produced after a suitable dwell time. The process has some similarities to the mixing of polyester resins for hand lay-up fibreglass but generally urethane plastics are used.

Each of these basic methods can be adapted and/or modified to suit particular requirements. For example, injection moulding machines have been designed with capstan-type heads carrying four or more differing moulds which are offered to the screw in rotation. This could have some significance for body panel manufacture. Also, it is now possible to compression-mould certain thermoplastic sheet materials, whereas thermosetting materials have had to be used previously. Similarly, it is now possible to

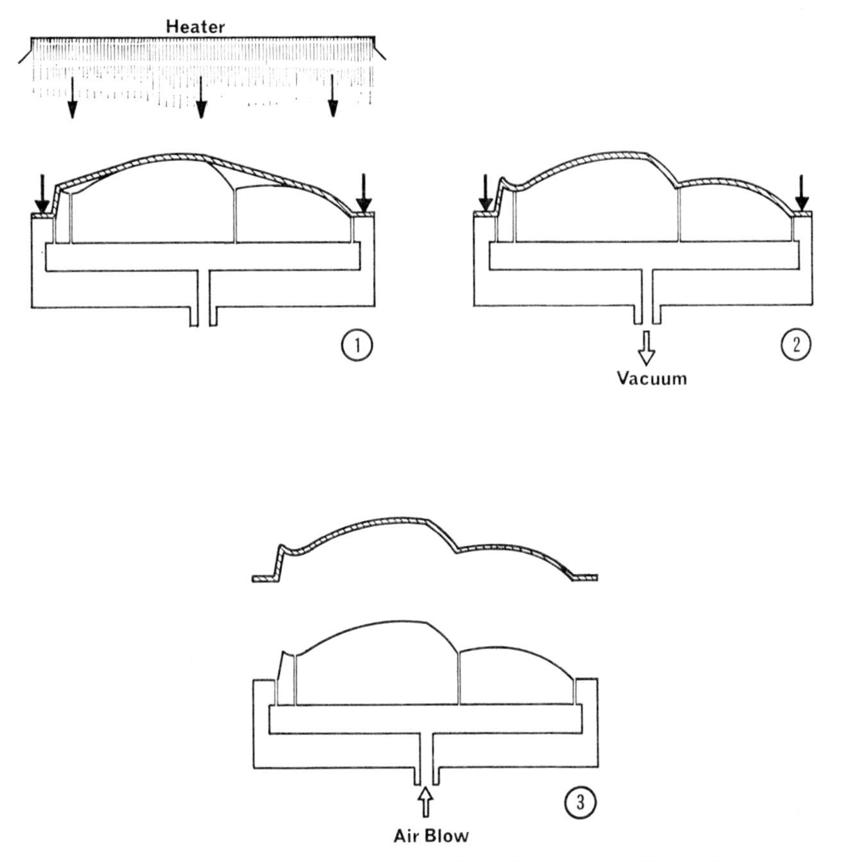

Fig. 4.5. In vacuum forming the sheet, in a flexible condition, is drawn tightly over the forming tool

injection-mould thermosetting materials. When the motor industry finally drops steel in favour of plastics for car bodies, the process will certainly become more continuous, if not automated.

Disadvantages

(i) Comparatively slow production rate.
(ii) Large floor areas and multiple moulds required.
(iii) Raw material expensive.

Advantages

(i) Possibility of large one-piece body structures.
(ii) Inbuilt stiffening ribs and bosses, with integral foam for added stiffness.
(iii) Low tool-costs.
(iv) Low equipment-costs.

MATERIALS

These various processes and the materials used in them are probably not in competition as it is more than likely that all of them will be complementary in manufacturing the all-plastics car-body of tomorrow; as indeed they already are in the manufacture of other components of motor cars today.

The material chosen as the most suitable for end-use will often govern the type of process to be used. For instance, a panel likely to require stiffness at prolonged high temperature would most probably be made from reinforced thermosetting materials. If it was large, compression moulding would be the process most likely to be used. A simple hang-on panel, like a boot lid, on the other hand requires no great heat-stability and only fair strength (except for impact properties at low ambient temperatures). The material here could be a thermoplastic with integral stiffening ribs produced by the injection moulding process.

This, of course, is the tremendous advantage of using plastics, when compared with steel. You can vary the materials and processes strictly in accordance with the end-use requirements.

The following present-day materials are suitable for automobile body and trim manufacture.

Polypropylenes (thermoplastics)

The range of materials in the polyolefins group has had a spectacular growth rate in automobile manufacture in the last five years. There is no question that polypropylenes offer the best range of all-round materials for a very wide variety of automobile applications. They are available at

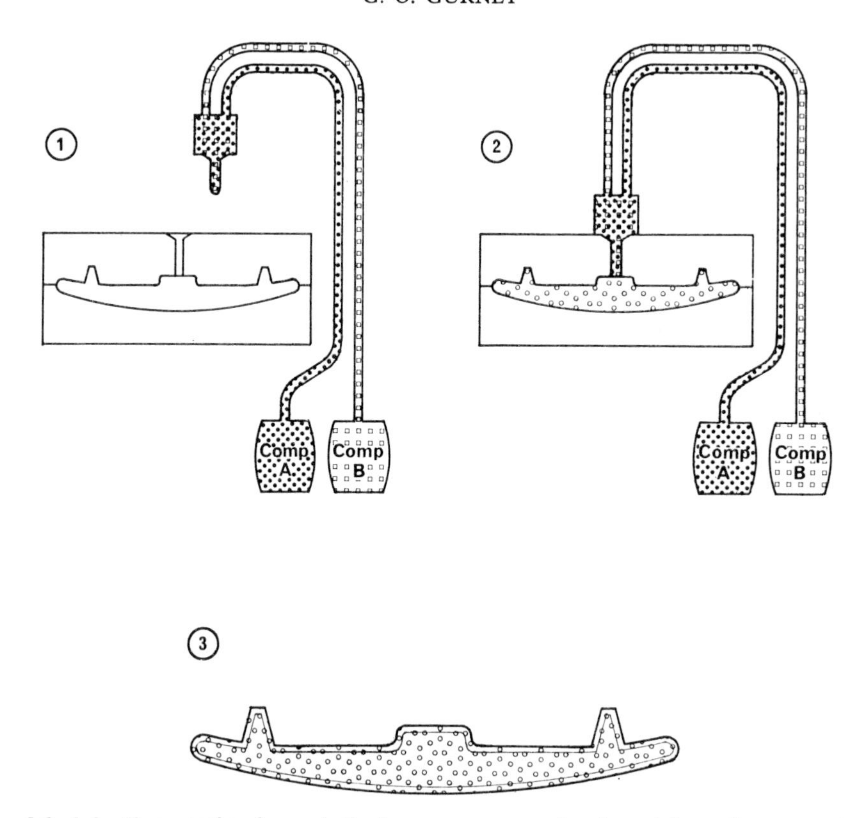

Fig. 4.6. Injection casting is a relatively new process developed from foam moulding

very attractive prices and the overall physical and mechanical properties of these materials cover most of the automobile industry's manufacturing requirements.

Long-term price forecasts indicate that higher tonnage could show truly significant price falls from the already good price levels. These materials are already being used widely for heater and fresh-air ducting, consoles, radiator fans and cowls, carburettor air intake boxes, facia panels, door pockets and, more recently in the U.S.A., fairly simple radiator grilles. Plateable grades of polypropylene are now available. Fibre-reinforced grades now are also widely available. Painting difficulties have not yet been entirely solved for exterior body panels but work is proceeding on this.

A word of warning to engineers on the use of polypropylenes: processing is not easy. These materials have fairly high shrink characteristics and are somewhat unpredictable in their after-moulding behaviour. Unexpected dimensional variations and distortion often show up on panels of complicated shapes.

The materials are suitable for fabrication by injection moulding, vacuum forming, rotational casting or blow moulding. Polypropylenes are of course well-known for their ability to form an integral hinge.

A.B.S. (Acrylonitrile-butadiene-styrene)

This is another range of materials which have also shown considerable growth in the automobile field in the last five years. They offer good stiffness and impact strengths, even at the low temperatures found in the Polar regions. They have fairly straightforward processing characteristics for injection moulding with excellent dimensional stability and reproducibility. They are perhaps best known for their spectacular use in the U.S.A. as radiator grilles. The most advanced European grille application is the one-piece grille and headlamp-surround on the Hillman *Avenger* GL. Like polypropylenes, A.B.S. have some potential as materials for external body parts, though this is limited by a softening point lower than that for polypropylenes and, at present, a somewhat higher price. However, the use of injection-moulded boot, bonnet and wings on Honda mini-cars and trucks has stimulated interest for these are high-volume vehicles by U.K. standards, at about 4000 per week.

A.B.S. materials are suitable for fabrication by injection-moulding, vacuum-forming and rotational casting. Blends of A.B.S./P.V.C. and A.B.S./Polycarbonate alloys are commercially available, the latter proving perhaps the most suitable for vacuum-formed body panels.

Painting problems on A.B.S. appear less significant than on polypropylene.

It is not yet clear whether Honda are fully satisfied with the technical performance of their body panels, but no particular problems appear to have presented themselves in service.

Nylons

These are excellent materials but, unfortunately, relatively high prices exclude their broad adoption, except for small, specialized components. However, I mention nylons because, without additional coatings, they will meet non-permeability requirements of fuel tanks.

Legislation, both in the U.K. and the U.S.A., is somewhat conflicting and unrealistic with regard to plastic fuel tanks, but in both countries interested parties are studying the question and it is thought that various nylon injection moulding or monomer casting grades will be competing with other plastics for fuel-tank production. Plastic tanks have better impact safety features and yet offer the prospect of full economic competitiveness with steel tanks.

Nylons are usually injection-moulded, although they may also be rotationally cast or blow-moulded. Glass-fibre-reinforced grades are available with very good mechanical properties.

A casting grade of nylon is available which can be polymerized in simple tools by adding a catalyst to the basic monomer before casting.

Fibre-reinforced polyesters

Fibreglass-reinforced polyester resins have already established themselves throughout the world as body materials. In few places, however, have they been processed by other than hand lay-up methods for body panels. The exception here is the U.S.A., where every major motor manufacturer has had experience in the field of F.R.P. panels manufactured by compression-moulding.

The Chevrolet *Corvette* has been manufactured in this manner since 1964 and the present *Stingray* version has a production rate of about 1000 per week. Most manufacturers have at least one model with an F.R.P. bonnet. Usually this is a high-performance model where fluted-top designs give a sporting appearance to an otherwise stock body.

Simultaneously, of course, this use gives the motor manufacturers an opportunity of evaluating this material in service. Considering its long period of use, it is surprising how little is known in this country about press-moulded fibreglass. Polyester resins usually have high shrinkage which causes painting problems as fibres stand proud of the surface and in the long term waviness appears. Recent developments have eliminated this problem and normal metal-body finishing techniques can generally be used.

This F.R.P. has possibly the most suitable mechanical properties of any of the plastics we could consider for body building. Cheaper sisal reinforcement can be used where strength and appearance are less critical.

Although usually processed by compression-moulding, fibreglass-reinforced polyester dough moulding compounds can also be injection-moulded but no large components have yet been made in this way.

Polyurethanes

This is a comparatively new range of materials about which little is generally known, as compared with the other plastics. They range from resins for producing plastic foam to injection moulding thermoplastic grades. Generally polyurethanes have not been considered as serious contenders for car body manufacture but recently grades which can be injection-cast have been developed. These have been the subject of considerable study by British Leyland, I.C.I. and others.

A special feature of injection-cast urethanes is that metal reinforcement-panels can be incorporated and fibre reinforcements can also be accommodated.

There is no point in presenting tables of properties of the various competing plastics as these are readily available from the manufacturers and from several well-illustrated volumes on the subject. Suffice it to say that the range of materials is very wide and the various properties can vary considerably.

This perhaps makes it very difficult for the automobile engineer to decide which material he should choose for a given component. A sound practice is for the design engineer to establish clearly the operational requirements together with the total envisaged production and then to let manufacturing engineers with specialized knowledge of plastics and processes make the final recommendation, based on the best cost/performance ratio.

SOME SIGNIFICANT RECENT DEVELOPMENTS

Several developments of particular significance have emerged recently which could well advance the date when mass-produced plastic car bodies become a reality. Perhaps the most exciting is the subject of British Patent No. 1-156-217 by I.C.I. This relates to an invention for the production of injection-moulded panels having a foamed core but with solid surface skins.

One of the disadvantages of the normal injection moulding technique is that ribs and bosses invariably 'grin-through' to the surface of the moulding. This is particularly evident in those materials which appeal economically to the motor manufacturer, such as polypropylene. However, the presence of a foamed core in an injection-moulded panel effectively removes the sinkage problem and allows quite large stiffening ribs and bosses to be incorporated without detriment to the surface appearance of the other side.

Furthermore, this patented process is capable of producing foam-core panels in a variety of differing materials with, consequently, different properties. Thus, a panel with plasticized P.V.C. skin and P.V.C. foam would prove very flexible, but a panel with glass-reinforced nylon skins and rigid urethane foam would provide a very stiff and strong structure.

This new process will perhaps overcome the basic disadvantage attached to thermoplastic materials for car-body manufacture, that is the unacceptable drop in stiffness at higher temperatures, without resorting to over-thick, and therefore over-expensive panels.

Whilst considerable development work has yet to be

carried out by I.C.I. to make the process suitable for very large panels, there is no doubt that it will be successful and its impact on the moulding of components for the automobile industry must be important.

Another significant development is in the competing thermosetting field and relates specifically to fibre-reinforced polyester resins shaped by high-output compression-moulding. Here again reinforcing ribs and bosses on large panels did, until recently, result in 'grinning-through' due to shrinkage during the cure cycle. Another problem caused by excessive shrinkage and a high thermal expansion was that the reinforcing fibre pattern became evident on the panel during normal painting. To get rid of this pattern and the sink marks meant an extra finishing operation.

Several manufacturers now offer what are termed 'low-profile' resins with virtually no shrinkage and much lower expansion coefficients. These new resins have made it possible for glassfibre-reinforced polyester body panels to be produced with inbuilt stiffening ribs which can be put through the normal painting plant of motor manufacturers without any additional surface treatments.

Coupled with this is the development of fibreglass-polyester sheet moulding-compound: low-profile resin-compound and cropped glassfibres are offered up together on a flat conveyer between two sheets of film. They are then compacted and wound up in rolls, ready for the moulding presses.

Whilst sheet moulding-compound has been available in this country for some time now and is often referred to as 'prepreg', the significance of recent developments is that the plant for manufacturing it is now available to moulders, so that raw material costs can be reduced. For any manufacturer contemplating moulding automobile panels from compression-moulded glassfibre/polyester, the initial cost of setting up a sheet moulding compound production line would be very quickly recovered by the low raw material costs achieved.

ECONOMICS

The two previous sections dealt with the various competing materials and processes which are often interrelated. The decision as to which is the correct material and process is an economic one. It should be based on the cost/performance ratio obtained from a detailed analysis of all factors.

In the case of plastics body-panel manufacture, studies carried out by a number of companies, both inside and outside the motor industry, have shown that the subject is an extremely complex one. Automobile engineers or senior executives cannot, at present, expect the firm cost guide-lines for plastic bodies that have been developed for metal-body manufacture. After all, firm guide-lines were not available at the time all-steel bodies were introduced. Such guide-lines will be developed as experience grows.

A similar situation obtains on the design front; the current unitary body construction has been developed to supersede the chassis types as a result of increasing confidence and design data availability: so it will be with plastics! Design adapted to the material will considerably improve the economic viability of the plastics car body as knowledge and confidence grows.

Of course, quite accurate costs for either individual panels or complete car bodies can be assembled, provided adequate disciplines are followed. It is, however, essential that all the following factors are properly considered and analysed if a realistic economic assessment is to be made.

Type of vehicle

A sophisticated sports car or city car would almost certainly show more favourable economics in plastics as against steel, than would a simple van body.

Design

A vehicle designed for plastics would show greatly improved manufacturing costs. For example, the clam-type structure of the Cycolac research vehicles or the Bond Bug type design yield better economics than the conventional saloon, which is strictly designed for steel.

Production rates

Each of the various competing processes has an optimum economic production rate. Injection-moulding produces an average of one panel every minute. A tool-change on a big machine will take all of one shift and more. It follows that, if satisfactory utilization of these expensive machines is to be achieved, then tool-changes must be minimal and component output must be very high or very large finished-part stocks must be accepted.

At the other end of the scale, vacuum-forming is a much slower process but tool-changes are very quick.

Tooling costs

Tooling costs are amortized against the unit cost so the higher the output, the higher the tooling cost that can be absorbed, and vice versa. High tooling costs, as for injection moulding, give the lowest unit cost for volume production but involve the highest financial risk, should the component or the car model fail in the market.

CONCLUSIONS AND THE EXECUTIVE DECISION

The executive decision to go to plastics or not, is not an easy one. Sufficient information is now available from many sources which proves that the economics is favourable for medium production runs of plastic bodies, at volumes hovering around 1000 per week.

There is also much evidence to show that plastics body manufacture at 2500+ per week volumes will become economically desirable before the turn of the century.

Most British manufacturers, however, have invested vast sums of money in modern metal-body building facilities and any immediate changeover would result in under-utilization of these valuable plants. Therefore it is not envisaged that plastics body facilities will become widely available until such time as the existing metal facilities are due for replacement.

With the ever-increasing demands on capital in the motor industry, any reduction in plant investment, and in new-model tooling and development costs, must be welcomed in the board-rooms. Plastics body manufacture can offer this reduction, thus substantially reducing the very high risks in decision-making on new projects. This risk is now so high that, even with modern management techniques, some board-room decisions can still mean life or death to works or even a whole company.

Paper 5

WINDSCREENS OF THE FUTURE

S. E. Kay*

Some suggestions are made on the technological changes which may occur during the next 30 years or so to make available windscreens with improved safety and optical performance and to give a better environment within the car.

INTRODUCTION

THE OBJECT of this paper is to discuss the technical changes that might be expected to occur in windscreens of motor vehicles during the next three decades. These changes can be predicted with considerable certainty from what we already know of the rate of technological exploitation of commercially viable products and processes; and from an estimate of the future rate of exploitation.

I will confine my remarks to technology which is firmly based. Excursions into the 21st century will be avoided. You will be aware that techniques are being developed for really long-term technological forecasting, but I think this subject is outside the scope of this symposium.

In order to discuss the future of the windscreen I have made two assumptions.

(1) That motor cars will continue to be the main form of personal transport during the period under consideration. The growth of the motor industry is obviously very dependent upon the quality and mileage of our road system and, as this becomes more nearly saturated, other forms of transport will have to be developed. I would not expect, however, that the growth rate of the motor industry would be affected significantly in the period under discussion by alternative methods of transport which require a different approach to vehicle design, for example, by tunnels, air flight or gas cushion support.

(2) That the driver and passengers will need to see out of the car for navigation purposes and personal comfort. Computerized traffic control of, for example, major inter-city roads and electronic aids to navigation and accident prevention are likely to lead to larger glazed areas than are in use at the present time.

The windscreen as we now know it, and as I believe we will continue to know it for many years to come, has to meet four prime requirements; or, as I believe it is expressed in body engineering terms, has four 'fixed points'. The first is to be transparent and scratch-resistant, the second to provide protection against the climate, the third is, when broken, to exhibit a type of fracture that will minimize injury to occupants or pedestrians; and the fourth is to maximize cost effectiveness.

STANDARDS

It might be worthwhile to discuss the quality of present safety-glass since this is the minimum standard for the windscreens which will be developed. The product must comply with the standards established by all national authorities. These are continuously being improved and the safety-glass industry has an important responsibility to develop the technology to permit these improvements. Present standards define optical requirements (particularly in the primary vision area of the driver), fracture characteristics and strength. In the U.K. the best known specification is BS 857—*Safety Glass for Land Transport*. Standards exist throughout the world, but they are not identical. All standards must be met in order to satisfy the export requirements of the British motor industry.

The British Standard optical requirement for a windscreen is that 8·4 minutes of arc primary deviation and 15 minutes secondary image separation shall not be exceeded at the windscreen mounting angle. These limiting values restrict the shape of the windscreen in the primary vision area and must obviously be a restraint on the freedom of stylists. For example, the radii in the shoulder area should not be less than about 400 mm and the radius of curvature in a vertical direction should not be less than

The MS. of this paper was received at the Institution on 21st April 1970 and accepted for publication on 1st June 1970.

* *Technical Director, Triplex Safety Glass Co. Ltd, King's Norton, Birmingham B38 8SR.*

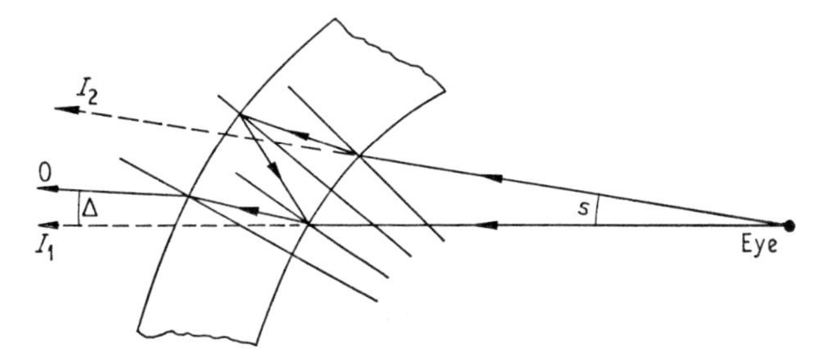

Fig. 5.1. The formation of primary and secondary images, seen through an inclined, curved glass screen

about 2500 mm for 50° rake and about 3000 mm for 55° rake.

The formation of the primary and secondary images seen through an inclined, curved glass is clearly illustrated in the ray diagram (Fig. 5.1). The changes in primary deviation and secondary image separation caused by changes in curvature of the windscreen are graphically illustrated in Figs 5.2 and 5.3, respectively, which show comparative values for three different windscreen shapes.

The greatly increased values of deviation and secondary image in the shoulder regions of a wrap-round windscreen are most marked on the passenger's side, though in the case of screen III with a graduated curvature, the peak values occur further out from the car centre-line, thus rendering this design optically more acceptable.

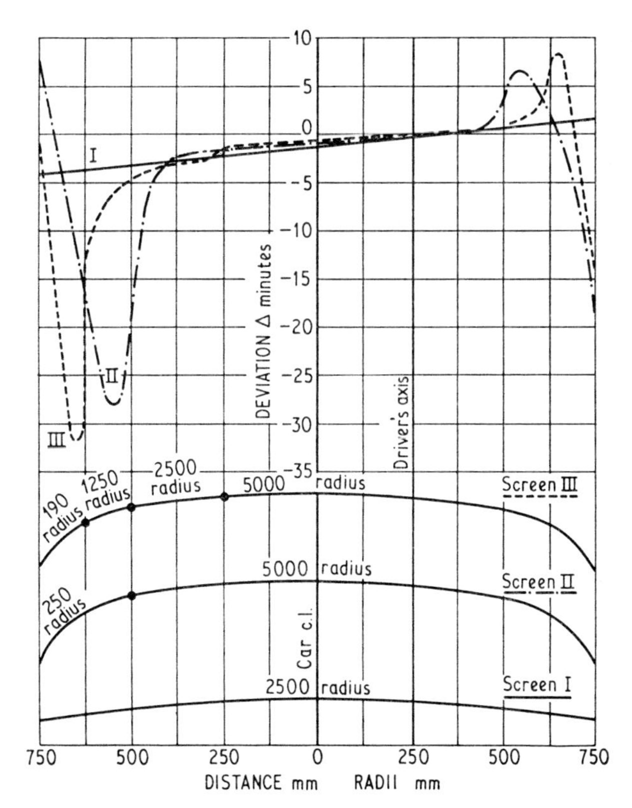

Fig. 5.2. The primary deviation caused by changes in curvature of the windscreen

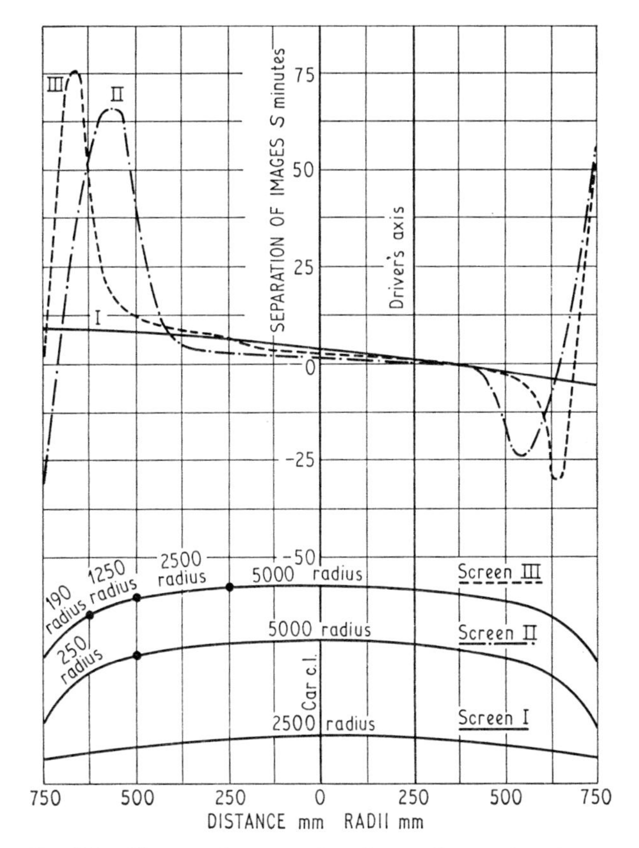

Fig. 5.3. Changes in secondary image formation due to the curvature of the windscreen

The optical characteristics of any new design of screen may be readily determined by means of the nomogram in Fig. 5.4, which shows the relationship between the relevant design parameters.

I should like to say something now about the strength of windscreens. This is measured in terms of the resistance to the impact, and penetration, of free-drop missiles; standard test missiles in common use include a 5 lb steel ball, a 22 lb form for simulated head impact tests, and a ½ lb steel ball. The injury potential of safety glass, when broken, is assessed either by the size and shape of particles after fracture in the case of toughened glass, or the penetration-resistance, laceration characteristics and degree of surface-splintering for laminated glass.

SAFETY

The assessment of injury potential of laminated glass is at present being studied in considerable detail by industry, universities, Government research laboratories and medical research groups. From this work it is to be expected that new, and substantially better, safety-glasses will be developed with the objective of a much improved performance.

Laminated glass windscreens are compulsory in cars exported to North America and certain other countries. There appears to be a slow but steady trend towards the

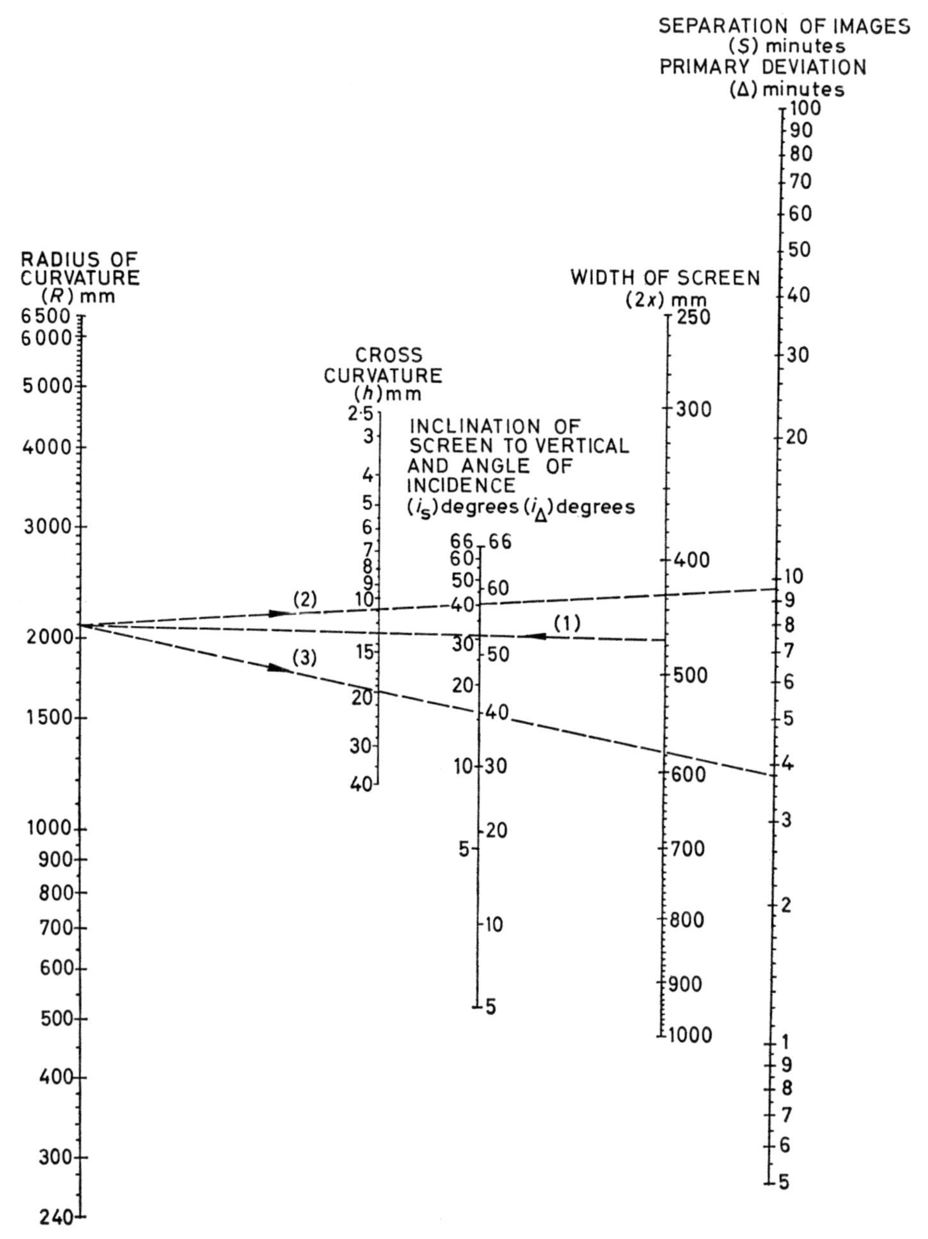

Fig. 5.4. This nomogram of the optical design parameters shows the characteristics of any new design of windscreen

more widespread acceptance of this product. This is because studies under simulated car crash conditions and analysis of road accidents have led to an understanding, especially in the U.S.A., of the importance of retaining the occupants in the car during and immediately following an accident.

Conventional monolithic toughened glass, stressed to the levels required for adequate visibility after fracture, will not meet this requirement. It is also unlikely that 100 per cent use of safety belts or 100 per cent reliability of alternative occupant restraint devices can ever be achieved. The general consensus of informed opinion is therefore towards a laminated windscreen with a satisfactory penetration resistance, satisfactory biomechanical performance (particularly for brain injury and laceration) and good service life. This promises to meet the requirements of the motor industry, motorists and pedestrians alike and to reduce the number of injuries from windscreens.

One of the problems is to define success and a variety of methods is being used to measure the performance of new products. Some of the most useful impact tests have been carried out with cadavers or dummies on sledges, with instruments embedded in their skulls to measure

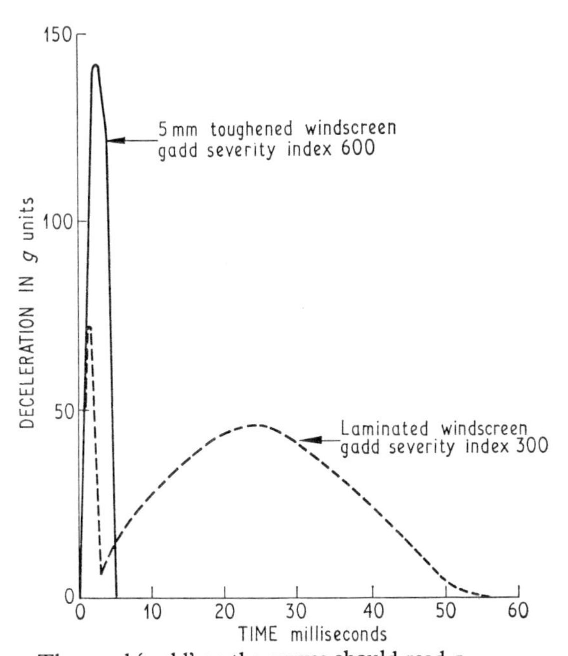

The word 'gadd' on the curves should read g_{add}.

Fig. 5.5. Typical deceleration curves for present toughened and laminated windscreens

deceleration forces and other parameters. Important information is also being gathered and correlated following on-the-spot accident investigations.

Severe injury or death can result from the compressive loads on the skull causing skull fracture and also from relatively prolonged high deceleration causing permanent damage to the brain. With the aid of this knowledge, severity indices have been defined for head-impact on windscreens. One of the best known of these is the Gadd Index which is the weighted area under a deceleration–time curve, or $\int_{t_1}^{t_2} a^{2\cdot 5}\, dt$. It has an upper critical limit set at 1000; $(t_2 - t_1)$ is the duration in seconds of deceleration and 'a' is deceleration in 'g' units. Typical deceleration curves for present laminated and toughened windscreens are shown in Fig. 5.5.

A further hazard of head-impact is laceration, either from the glass at the point of impact or from that retained in the glazing channel; or, more severely but now less likely, penetration of a laminated windscreen. The former could result in disfigurement while the latter could prove fatal due to cuts. The development of penetration-resistant interlayers may also be expected to reduce injuries. The indications are that the use of the present conventional laminate, which includes a thicker interlayer, has already resulted in reduced fatality and injury rates. There is also a less publicized, but nevertheless important, problem of pedestrians being injured when thrown up over the bonnet and into the windscreen. Windscreens must therefore also provide protection against laceration and head impact damage when struck from either side.

You will have understood from all this that the windscreen is a complex component of the motor car.

REQUIREMENTS

Let us now examine the main factors which can be expected to generate changes in windscreen design.

Safety

Firstly there is safety. There is evidence that national and medical requirements will lead to a continuous pressure for new windscreens to be developed which have improved safety performance under all motoring conditions. Society will insist on products which will reduce personal injury and will be willing to pay for them. Legislation elsewhere, for example in the U.S.A., will obviously pose requirements for cars manufactured in the U.K., if only to maintain exports.

Already we can see evidence of new types of experimental windscreens emerging from many safety-glass manufacturers with the object of improved safety performance. Some of these include glass of a special composition which has been subjected to specially developed processing conditions. Other ideas are based on glass of different thicknesses, different moduli of rupture and so on. There are ideas based on composites with a glass outer and an organic sheet inner. Much work is also going on in the field of new types of adhesive interlayers. As well as weighing less, most of these new windscreens have a better safety performance.

Road resistance

Windscreens will be required to have a long service life with retention of visibility after collision with flying road particles. The development of improved surfacing materials and methods might reduce the incidence of impact from these on major roads but our present problems will remain on minor roads, and also, of course, on roads in less developed countries. New road surfaces may affect our present views on vibration, and adhesive attachments may become commonplace.

Comfort

Heat and glare-reducing properties, probably achieved by special coatings, will be particularly important when large areas of the car are glazed, as will be the development of effective demisting and de-icing techniques, most probably by electroconductive methods. In all these fields there is great activity. Standards for acceptable levels of polarized pattern might also be introduced for windscreens containing stressed members.

Vehicle performance

The biggest contribution to improved vehicle performance will be made by the use of thinner glass which will reduce overall weight and especially the weight above the vehicle's centre of gravity.

An improved performance of the windscreen itself in service will also be required. This might result from scratch-resistant coatings which can withstand the

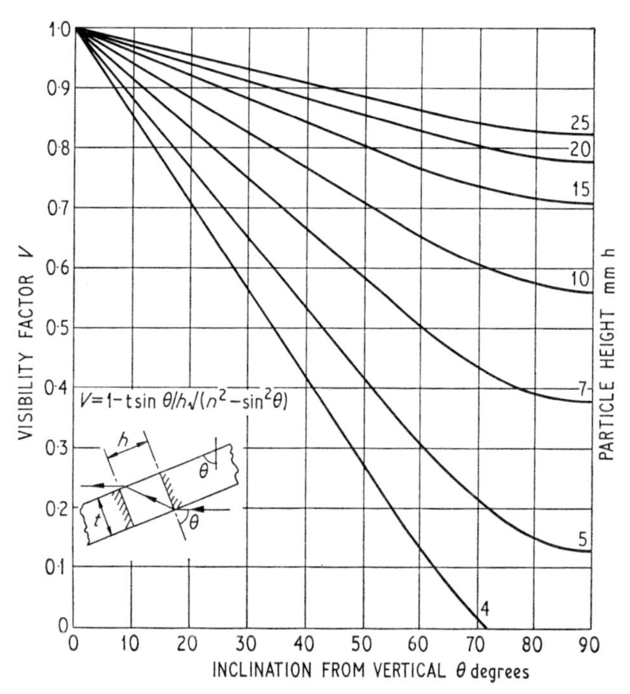

Fig. 5.6. The visibility factor depends on glass thickness, fragment size and screen mounting-angle

abrasive action of a windscreen wiper with road-grit adhering to it.

Styling

It is expected that computer-graphic design of car windscreens will become commonplace and this may lead to glass being considered as part of the mechanical structure in car body design. The development of glass with heat and glare-reducing properties will lead to large areas being glazed. With much improved strength characteristics, it will become possible for the windscreen to be close-bonded to glass roof panels and side glasses. The windscreen will then become a structural member, as it is now on aircraft flight decks. Styling trends might possibly separate into simple shapes for the city-type car and the more complex styling for the motorway car where aerodynamic features are more important.

The high standard of optical quality required, particularly in the primary vision area, will limit permissible rake angle and hence restrict the freedom of the stylist. A similar restriction is already imposed for toughened glass where, for rake angles above about 55°, the zoning requirements for visibility after fracture conflict with the safety-after-fracture requirements. Whereas visibility through a laminated windscreen may not be greatly impaired by fracture, considerable loss of vision will occur when a toughened windscreen is broken. The degree of vision after fracture may be expressed as a factor which depends on glass thickness, fragment size and screen mounting-angle, in accordance with Fig. 5.6.

Adhesive glazing techniques will probably become commonplace, not only for improved weather-resistance, but also to permit a continuous line between body-panels and glass for improved appearance and reduced wind noise.

Body engineering

As mentioned earlier, glass may be required to play an increasing role in the overall structure of the vehicle. Assuming that glass can be developed to meet these requirements, one can envisage the 'A' post being moved further back, or completely eliminated. Already the use of rollover bars permits a much greater use of safety-glass in body glazing. The structural role would also be increased by the use of high-strength fixed side glasses and this concept may become more important as improved air-conditioning reduces the need for opening windows. Direct, mechanical attachment of windscreens to the body shell may become widely used, either by pre-glazing into frames or by specially developed edge designs.

Environmental conditions

So far the windscreen has been considered only from the visibility and safety aspects. It could, however, be developed as a carrier for other components. Radio aerials can be incorporated; instrument displays are possible. Temperature or light-sensing elements could be included for automatic control of the heat and light which enters the car through the windscreen.

In the period with which I am concerned, I do not believe that we shall see photochromic glasses being developed to the level required for automotive use. They may possibly be used outside the vision area, e.g. for roof glasses, but the windscreen application demands response time, reliability and optical quality which probably exceed the likely technological developments.

Demands will be maintained for some new methods of keeping the outer surface of the glass clean under all likely conditions; and also of overcoming our national fog problem. However, I believe such developments will occur outside the glass industry and may follow from aerospace development.

CONCLUSIONS

In conclusion, the properties of a windscreen I see as being in common use in about 20 years' time can be summarized as follows:

(1) It will be a laminated windscreen, comprising three members; the outer certainly glass and the inner probably glass, each being about 2 mm thick, with an interlayer whose thickness and other properties will optimize safety and mechanical performance of the laminated composite.

(2) Its performance will be so much improved over the present laminate that head-to-glass impacts of up to 50 mile/h will cause only very slight lacerative injuries. Pedestrians will survive similar impacts.

(3) Visibility sufficient for all normal driving purposes will be maintained after impact with flint and road

stones, and windscreens will not be penetrated by these objects.

(4) Heat and glare-reducing characteristics will be included for improved personal comfort and this may lead to larger glazed areas.

(5) It will be electrically heated for demisting and de-icing by devices which may or may not also be heat and glare-reducing.

(6) It will be more resistant to abrasion.

(7) It will have great mechanical strength and form part of the structure of the body shell. It will probably be mechanically or adhesively bonded to the aperture.

Paper 6

BODY COMPONENTS

A. E. Norman*

INTRODUCTION

FIG. 6.1 shows the type of units which we call body components. All of these pose their own particular problems. The chosen solution must be that component which performs all the functions required, fits into the space available and can be manufactured at an economic price.

Probably the greatest problem is that posed by the door on modern European vehicles. Here, the number of components which must be fitted into a very confined space is considerable, i.e. hinges, door check, latches, handles, etc. We shall therefore concentrate on the door components.

DOOR COMPONENTS

Window regulators

Initially car doors were simple wooden structures, with a latch at one end and hinges at the other. With the advent of saloon cars, moving windows were introduced. The three most widely used systems are the drum and flexible wire type, which requires separate pulleys to be fitted within the door to guide the wire; the pinion and flexible rack type; and the pinion and segment type. The latter two are self-contained.

Development of window regulators is in the realm of power operation and the most widely used power source is the electric motor, which is shown on Fig. 6.2, coupled to a pinion and segment regulator.

The evolution of vehicle bodies with frameless glass windows has led to improvements in glass guiding systems. In the latest versions the glass is only tipped against the sealing rubber when it is virtually in the up position. This eliminates to a large extent the frictional drag on the glass.

The MS. of this paper was accepted for publication on 21st April 1970.

* *Wilmott and Breeden Ltd, Amington Road, Birmingham B25 8EW.*

Hinges

Progress has been made in the field of hinge design by making the door check an integral part of the hinge. This reduces the number of components which have to be fitted, and also the reinforcement required within door and pillar.

Latches and handles

The earliest type of latch was the slam bolt, developed from the domestic door latch. Attempts were made to prevent body distortion by incorporating a dovetail system in the latch as shown in Fig. 6.3. This was not completely effective since a resilient coupling was necessary between the striker and its mounting plate to accept tolerance variations between door and pillar. These latches were operated by a turn-handle from the outside and frequently by direct action of the latch from the inside.

Door latches became more sophisticated in the late 1940s and early 1950s. Latches were required to impart greater rigidity to a vehicle and at the same time to be capable of release by push-buttons. This latter was not feasible on slam bolts because of the release energy requirements and therefore another generation of latches called 'detent mechanisms' was evolved. The energy required to release such a latch is considerably less, as shown in Figs 6.4 and 6.5. Fig. 6.6 shows a selection of typical modern latches.

SAFETY LEGISLATION

Up to the middle of 1960 the specification of latches was left to the car manufacturer. This was radically changed by the introduction in America of legislation specifying functional and strength requirements for latches, based on recommendations by the American Society of Automotive Engineers.

These regulations, together with proposed regulations on side-impact resistance, means that the latch can no

Fig. 6.1. Typical body components

Fig. 6.2. Pinion-and-segment type power-operated window regulators

Fig. 6.3. This dovetail type of door latch was meant to prevent body distortion

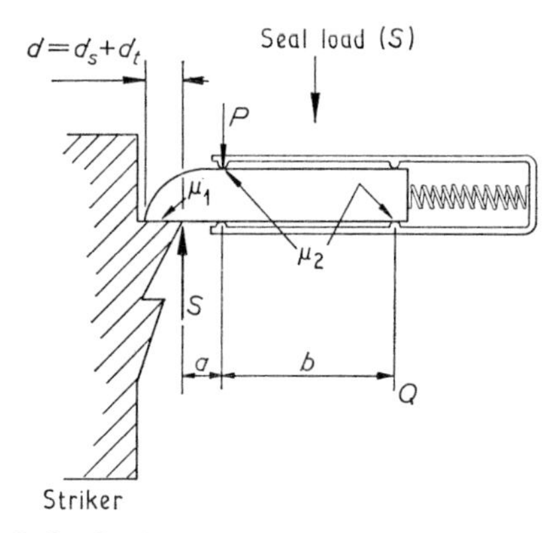

Fig. 6.4. Bolt displacement (d) of slam-bolt mechanism consists of d_s for strength and d_t for tolerances. Work done to withdrawn the bolt over d is $dS\left[\mu_1 + \mu_2\left(1 + 2\frac{a}{b}\right)\right]$

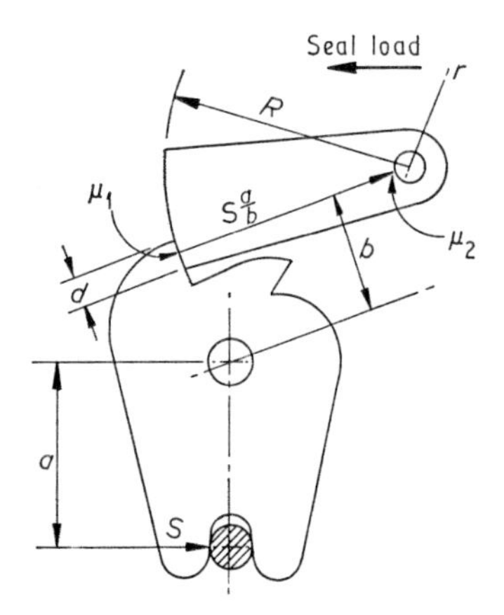

Fig. 6.5. Work done to withdraw the pawl of this claw latch is $dS\frac{a}{b}\left(\mu_1 + \mu_2\frac{r}{R}\right)$

longer be considered in isolation. We must now consider the complete assembly of latch, inside and outside release mechanisms and their attendant connecting linkages, as well as reinforcements within the door and pillar at the attachment points.

Fig. 6.7 shows a latch mechanism designed—when used with the appropriate release members—to meet the functional legislative requirements and considerably to exceed the strength requirements of current legislation.

Fig. 6.8 illustrates the way in which the inside release handle and window-winding handle have been modified by recessing the release handle and making the winding handle in flexible plastic, to meet requirements relating to projections within the vehicle.

PRODUCTION ECONOMICS

For economy the components must be in a form capable of production in very large quantities at a low piece price. Steel pressings are an example: these have been widely used and the design has centred on this material. Recently newer materials, such as plastics, have been used and economies can be achieved where the full range of the material properties can be utilized. The latch shown in Fig. 6.7 is an example in which the bearing qualities of non-metallic guiding faces and the flexible nature of the material have all been utilized.

Further economies could be achieved by much greater standardization. Whilst this has not been very successful in the past, it is possible that legislative requirements will encourage manufacturers in the wider use of proven, standardized mechanisms.

DESIGN TRENDS

In the field of innovation, one probable development is in the automation of conventional locking mechanisms by pneumatic or electrical power. With suitable circuitry a wide variety of control systems could be devised. For example all doors could be locked from a single switch on the facia, or it could be arranged that all doors automatically locked, once the vehicle exceeded a certain speed. It is probable that any such system would be additional, leaving the present manual controls fully operative in case of power failure.

A second possible innovation stems from increasing concern with side impact. The door stiffening member should ideally be as close to the outer panel as possible. This means that the inner panel could become superfluous from a strength aspect. An inner rail is possible which would carry not only the latch but also the window regulator and its attendant guide systems, as well as the inside release handle, with the possibility of the outer handle being connected by flexible cables. Such a unit is shown on Fig. 6.9.

Fig. 6.6. A selection of typical modern door latches

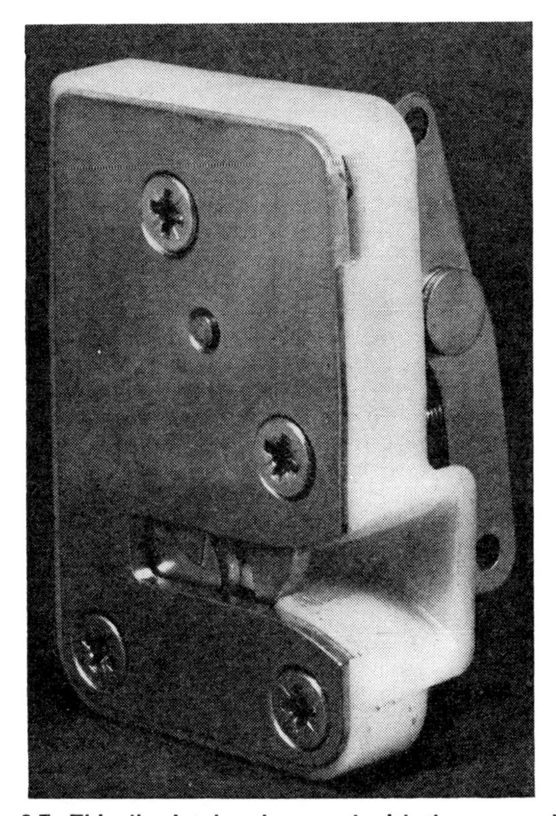

Fig. 6.7. This disc latch, when used with the appropriate release members, meets the legislative requirements and has considerable excess strength

Fig. 6.8. How the inside release and window handles have been modified to meet the safety requirements for projections within the vehicle

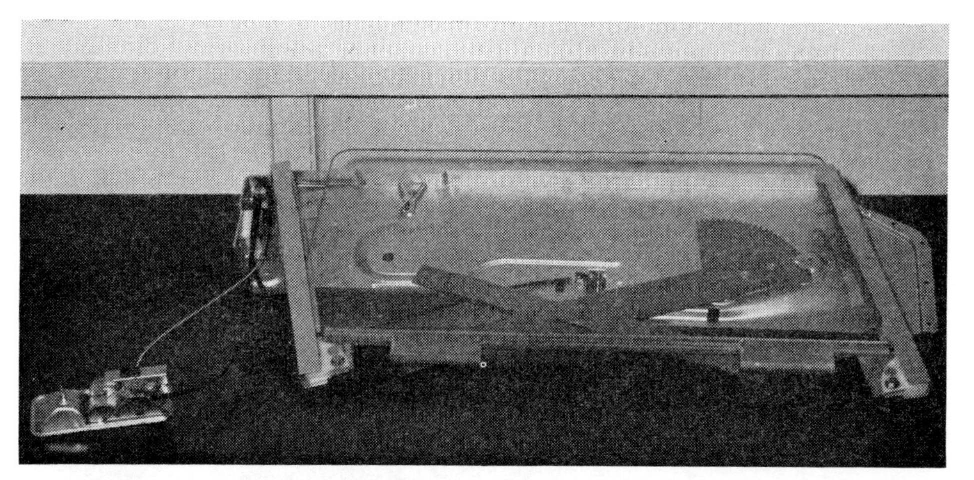

Fig. 6.9. Here the inner stiffening rail not only carries the latch but also the window regulator, its guide systems and the inside release handle

CONCLUSIONS

Components have evolved in a fairly logical manner together with the whole motor car, from fairly crude origins to present-day sophistication. In the future we can probably expect to see:

(1) A greater degree of standardization;

(2) An increase in sophistication, e.g. automation of the mechanisms;

(3) Very high-strength door latches;

(4) The use of new materials to improve function and give trouble-free service without attention throughout the vehicle's life.

Paper 7

STRUCTURAL ANALYSIS OF CAR BODY SHELLS USING COMPUTER TECHNIQUES

G. G. Moore*

To predict as many characteristics as possible of projected body designs, a computer programme has been developed on the basis of idealized beam and panel lay-outs. Results correlate well with actual tests but complete optimization is impossible because of the complexity of the structure and the number of different section properties. The methods of analysis are described.

INTRODUCTION

As competition between the major car manufacturers increases, so does the need for lighter and more efficient body structures. Until recently, the choice of section sizes and material gauges has been based somewhat arbitrarily on previous experience. Until the prototype body has been built and tested, it is not known whether it will be satisfactory.

By that time tooling can be at an advanced stage. Hence, if the structure is found to be too weak and modifications have to be made, costly increases in design time can occur. Conversely, if the body is too strong, the unnecessary increase in weight will be reflected in the cost of the body shell. Thus more knowledge of the structural behaviour of the body at an early stage in the design is very desirable.

Theoretical structural analysis by finite element displacement techniques has been well documented and proven, both in civil and aircraft engineering (1) (2)†. Earlier workers (3) (4) (5) had demonstrated that such methods can be applied to the analysis of car bodies. However, the idealization of the structure into beam and panel elements is much more difficult. Therefore the correlation with experimental displacements has not always been consistent and often the models have not responded to small modifications, a factor in which designers are primarily interested.

Several suitable computer programmes are now commercially available for basic structural analysis. Hence there is little point in expending further resources in this direction. Such a basic programme is being used by Pressed Steel Fisher Limited in the analysis of car bodies and commercial vehicles. Effort has been concentrated on ancillary programmes which help in the production and checking of data and the automatic analysis and plotting of results. This has enabled analysis to keep pace with the body designers and obtain preliminary results before the final drawings have been completed, while there is still time for modification.

These programmes have also enabled the problem of idealization, particularly with reference to joint stiffness, panel curvature, element size etc., to be investigated in general, with a view to achieving more consistent and accurate results. Once a satisfactory model is devised, any type of loading can be simulated. At present it is limited to the standard beam and torsion tests, certain suspension loads and those required by safety regulations.

A typical example is discussed in the paper and methods of successfully using the results are suggested.

METHOD

The method requires that the body shell be divided into an ideal three-dimensional framework, consisting of simple beams and shear panels, as shown in Fig. 7.1. The main elements with their associated properties are shown in Fig. 7.2. A rigid portion can be specified at the end of each beam. This facility can be used to account for finite joint-size and for eccentricity of beams meeting at a node. The ends of the beams can be either rigidly fixed, pinned or hinged. The panels are effectively pinned at the nodes to which they are connected.

The beams and panels are interconnected at nodes.

The MS. of this paper was received at the Institution on 8th May 1970 and accepted for publication on 27th November 1970.

* *Research and Development Dept, Pressed Steel Fisher Ltd, Oxford.*

† *References are given in Appendix 7.1.*

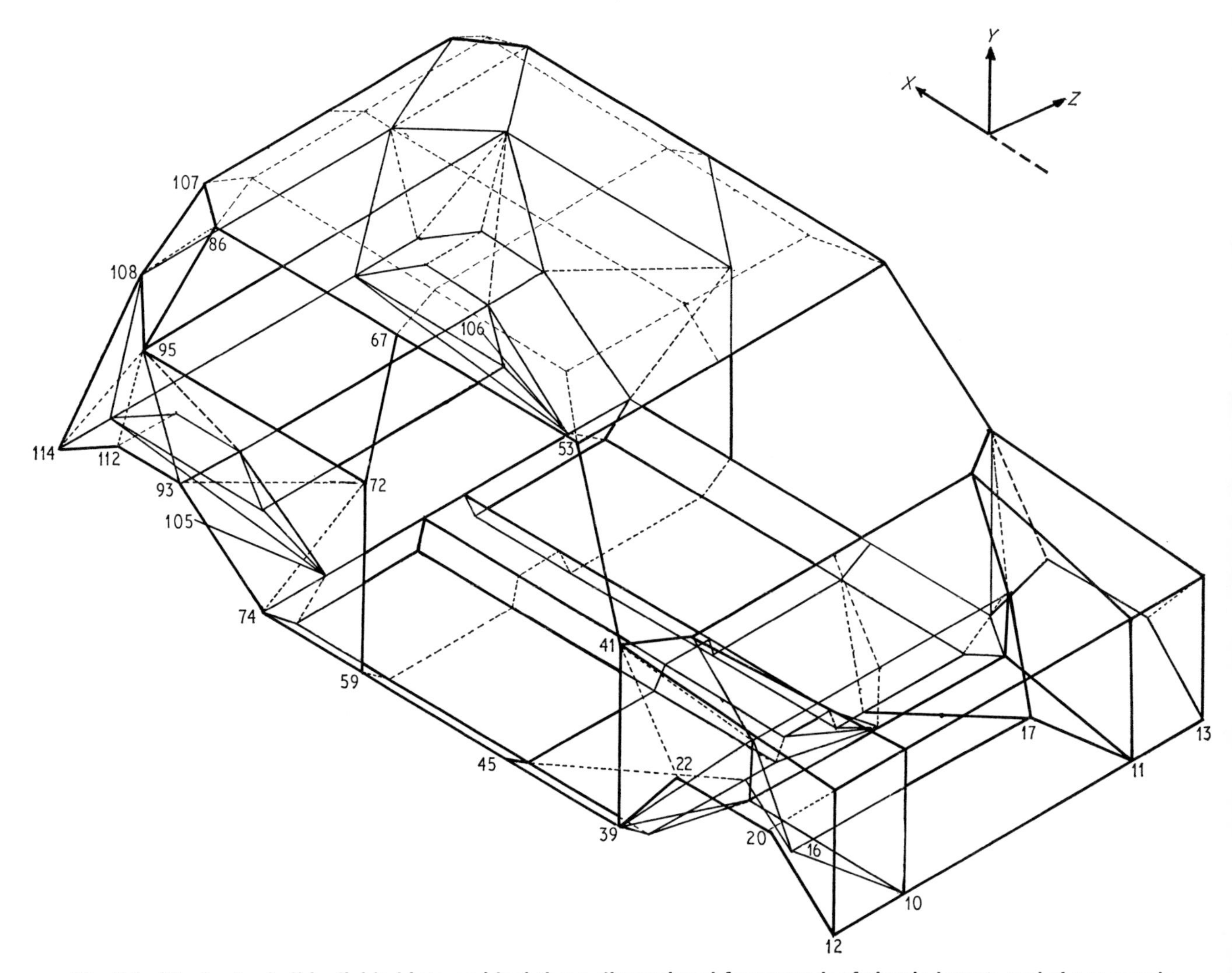

Fig. 7.1. The body shell is divided into an ideal three-dimensional framework of simple beams and shear panels

These nodes can be at any convenient points in the framework, usually at joints or changes in section. The geometry is defined by the node co-ordinates with reference to three arbitrary axes (x, y, z), and the beams and panels by the node numbers which they join.

Any loading system can be applied to the framework, which must be constrained at a sufficient number of nodes to keep it in equilibrium.

The method of solution, known as the 'stiffness method', is to obtain the equilibrium equations at each node (six per node, i.e. three forces and three moments) in terms of the node displacements (three deflections and three rotations). The node displacements are obtained by solving the equations. Individual panel and beam forces and bending moments are calculated by back-substitution of the displacements. Hence the resulting stresses can be calculated.

COMPUTER PROGRAMMES

The structural solution is obtained by a civil engineering programme which is now being run on an IBM 360/40 computer. The earlier results were obtained by using the facilities of various computer bureaux.

The input to the programme consists of node numbers and co-ordinates, member numbers and indices, member properties, constraints and load configurations. In addition, matrices of the stiffnesses of individual members can be direct inputs. This is a useful feature when, for example, specific joint stiffnesses or deep beams are under consideration. The output consists of node displacements, beam and panel forces and stresses, and beam moments.

A flow diagram showing the general procedure and computer programmes involved in the complete analysis of a car body shell is shown in Fig. 7.3.

The input data are obtained from layout schemes which can be on the drawing board. Nodes are fixed and co-ordinates defined with reference to the body 'ten' lines. At this stage any numbering system can be used which permits the nodes to be added or deleted at a later date. The main beams and panels are then numbered and defined

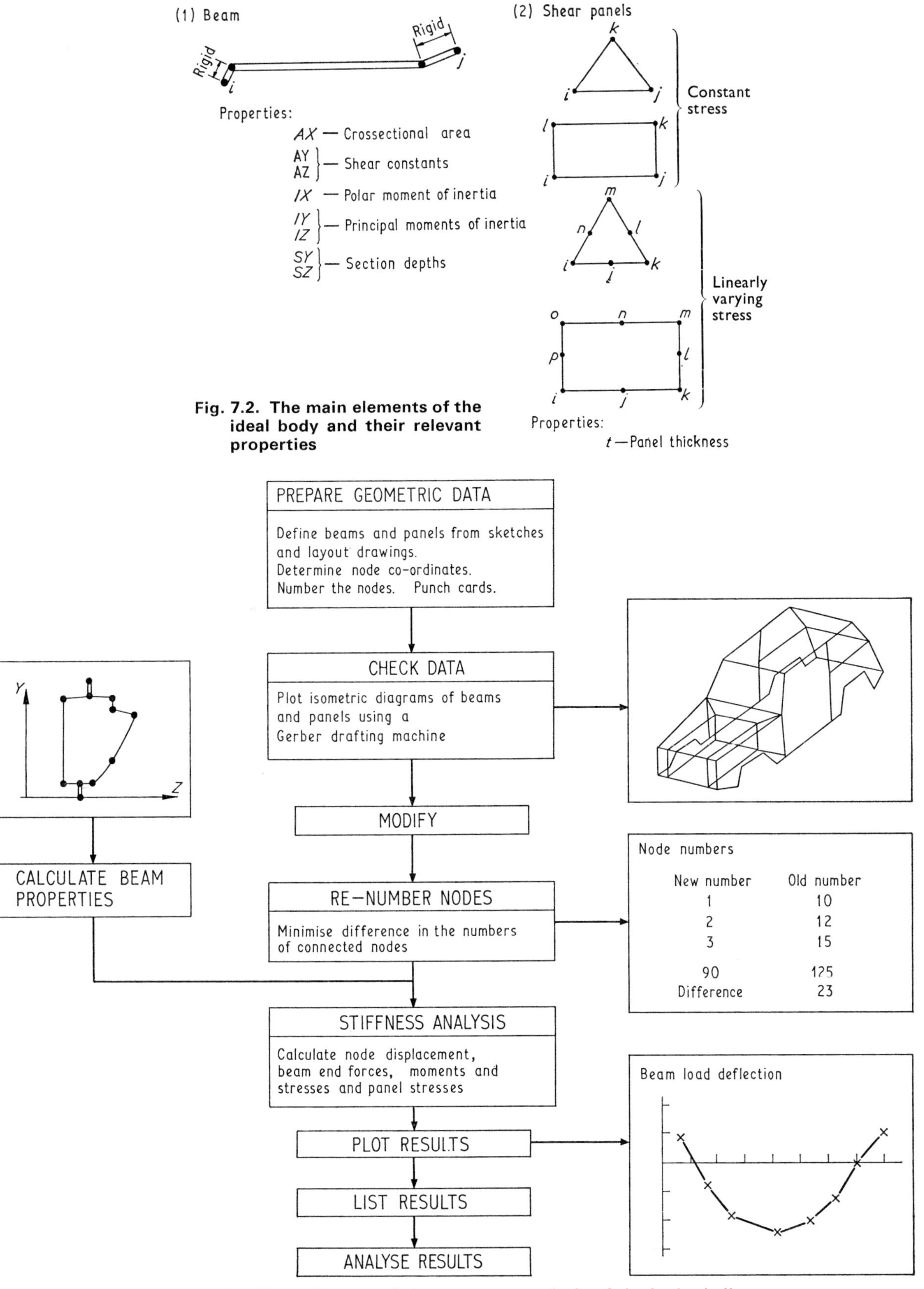

Fig. 7.2. The main elements of the ideal body and their relevant properties

Fig. 7.3. Flow diagram of the computer analysis of the body shell

by their node indices. These are punched on cards in a format which is suitable for input to the main structural analysis programme.

An isometric drawing of the complete framework is then produced from the punched data, using a Gerber drafting machine. A beam and panel drawing is usually done separately to provide a visual check on the geometry of the framework. The drawing can be reproduced at any convenient angle to give the clearest possible view and individual parts of the framework can also be reproduced. An IBM 2250 display unit is used in choosing the views.

The largest part of the computer time (about 75 per cent) is used for the solution of the node equilibrium equations and this becomes very important when considering large structures. It can be reduced very appreciably by 'banding' the stiffness matrix. This requires that the numbers of nodes on any beam and panel should be as similar as possible.

A simple programme has been written to re-number the nodes automatically. This starts off at Node 1 and re-numbers all the nodes connected to it consecutively, before moving on to Node 2. Other sophisticated means of re-numbering result in only marginal improvement.

The output from this programme is a listing of new and old node numbers. There is no need physically to re-number the nodes, it is only necessary that on the input the original node co-ordinate data should be listed in the revised numerical order.

The beam properties are worked out in a separate routine. They are usually modified according to the type of section (i.e. open or box) and joint conditions. Attempts are being made to reduce the tediousness of this by using a curve follower, e.g. the Gerber drafting machine, to get the section shape directly into the computer.

A considerable amount of information is obtained by this method, and the time it takes to analyse is reduced by plotting displacements along specified lines.

The time taken by the computer for the solution is approximately proportional to the cube of the number of nodes (N^3). A saving can be made on large structures

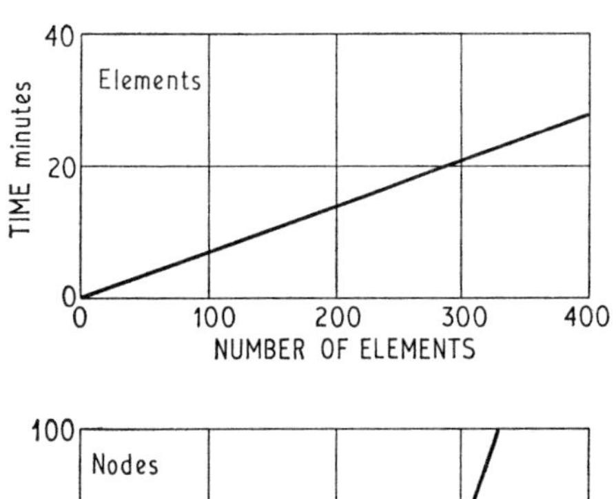

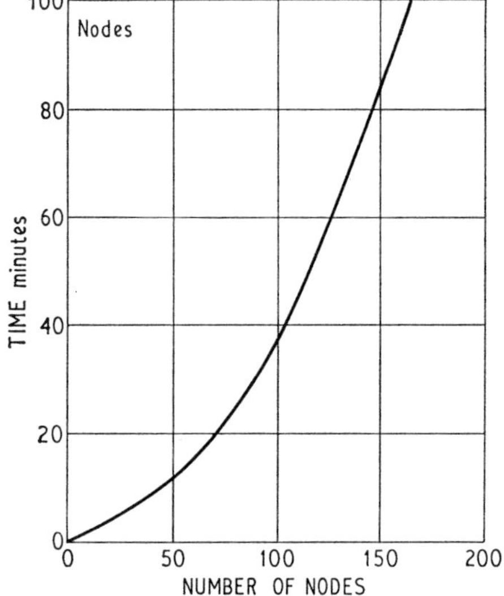

Fig. 7.4. These typical running times for solution on an IBM 360/40 indicate that considerable time savings are possible by this method

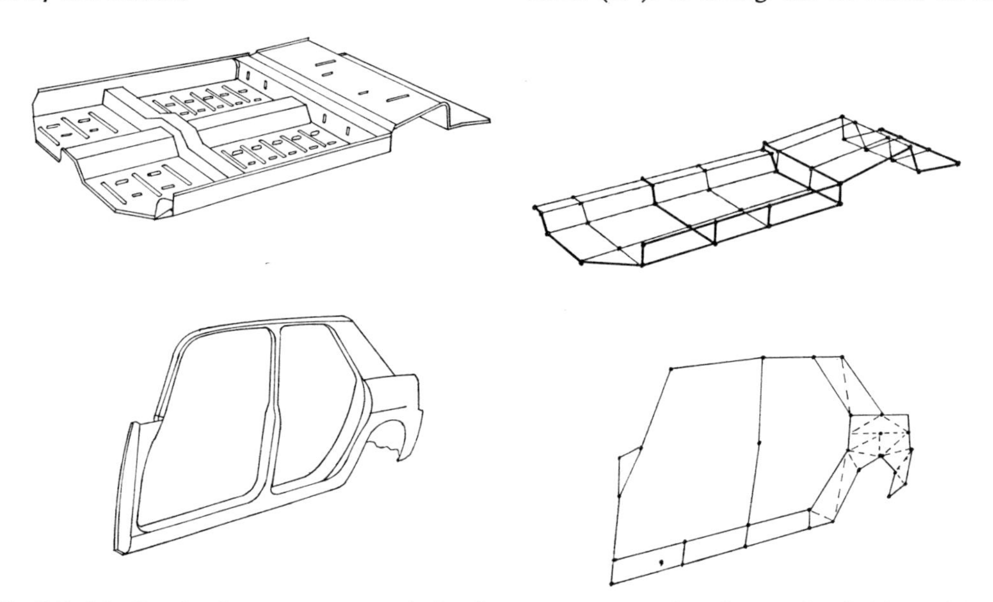

Fig. 7.5. Idealization for computer analysis of a transverse-engine, front-wheel-drive vehicle

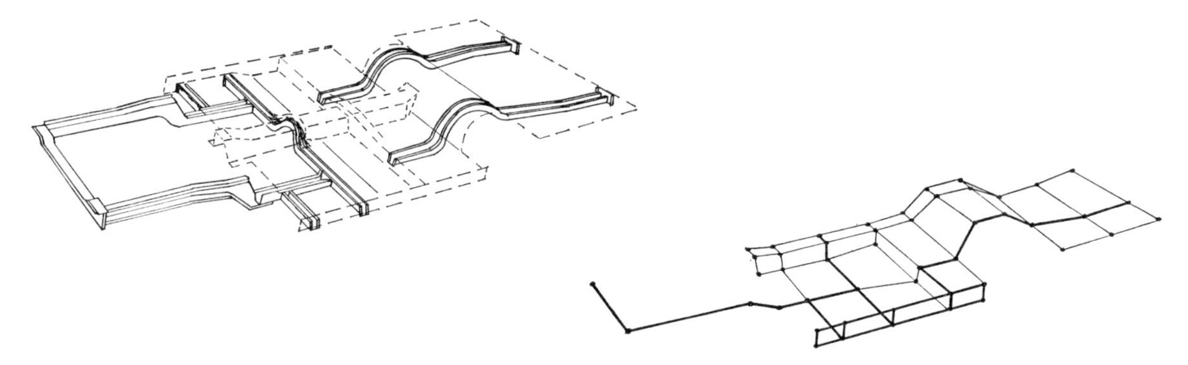

Fig. 7.6. This type of vehicle has a substantial floor structure with a deep tunnel

which are symmetric about a central plane. Only half the structure need be considered, artificial constraints being introduced at the nodes in the central plane. Unfortunately asymmetric and symmetric loading cases require different constraint conditions. Hence two separate runs are required for 'torsion' and 'beam' tests which still gives a time saving proportional to $2.(N/2)^3 = N^3/4$. This ideal is never quite achieved because further nodes are usually added in the plane of symmetry. Nevertheless, a considerable saving can still be achieved. Some typical running times are given in Fig. 7.4.

IDEALIZATION

Two basic types of body structure are made by the British Leyland Group: the transverse-engine/front-wheel-drive vehicle and the conventional longitudinal engine/rear-wheel-drive.

The first type usually has a fairly flat floor with a shallow tunnel and front-seat cross-member and comparatively stiff side frames, as shown in Fig. 7.5. The suspension or subframe mounting-points transmit the loads almost directly into the side frames. Good correlation between theory and experiment was usually obtained for beam tests and reasonable correlation for torsion tests. This indicates that the joints in the plane of the side frames are fairly efficient but the transverse connection between the side frames becomes critical in torsion.

The second type of vehicle has a substantial floor structure with front and rear longitudinals, front-seat cross-member and deep tunnel, as shown in Fig. 7.6. The side frames are not so substantial as on the front-wheel-drive vehicles. The suspension loads are fed into the passenger compartment through the longitudinals and the floor structure.

In both beam and torsion tests this type always appears to be stiffer in theory than it really is. It has been shown that a significant change in overall torsional stiffness can be obtained by varying the floor-joint conditions. In practice the joints are almost 100 per cent efficient in bending in the plane of the floor, but less efficient in torsion and very poor in bending outside the plane of the floor.

This applies particularly to the cross-member/sill joint shown in Fig. 7.7. Here the flanges of the cross-member are spot-welded to a flexible panel and offer very little rotational stiffness. Hence the cross-member is assumed to be hinged about the x-axis. Other joints can be simulated by modifying the beam properties and some are obviously 'pinned' because of their design (e.g. B/C post-cantrail joint).

Deep beams and their connection to the framework pose a problem, particularly the sill-heelboard connection. It can be overcome by dividing the sill into two beams, connected by a shear panel as shown in Fig. 7.8. The properties are chosen so that the overall stiffness of the beam–panel combination is the same as the original sill. Alternatively, the exact stiffness of the matrix can be calculated for direct input to the computer.

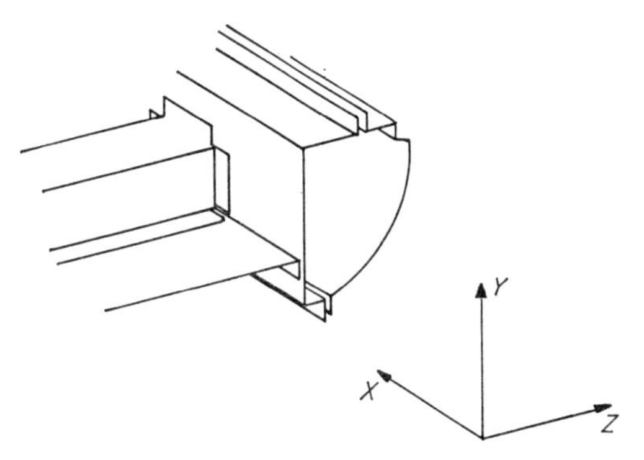

Fig. 7.7. The cross-member/sill joint completely transmits bending in the plane of the floor

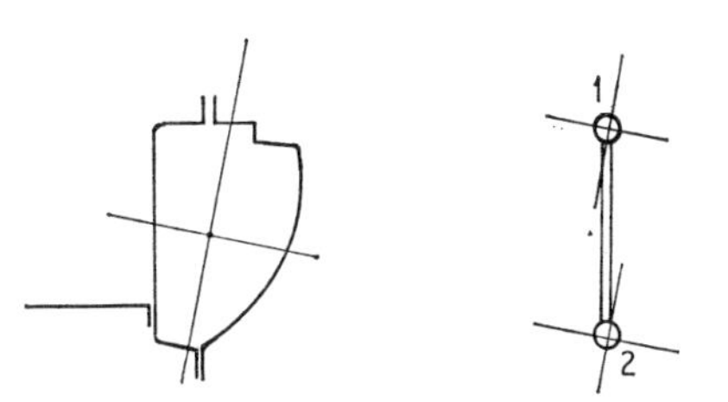

Fig. 7.8. The sill is divided into two beams connected by a shear panel

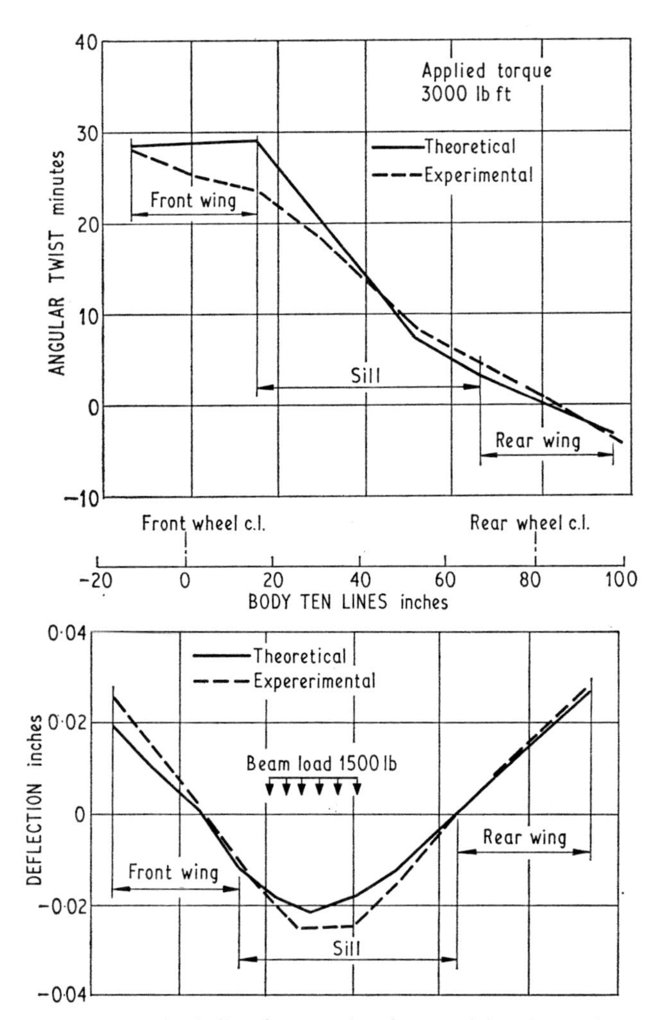

Fig. 7.9. Body deflections and twist resulting from the beam and torsion tests

Recently up to about 250 nodes have been found for the complete body and it does not seem like that 350 nodes will be exceeded. This number can quite often be reduced by careful inspection, but the amount of computer time saved does not warrant the extra time spent on the idealization.

RESULTS

Typical example

The idealization shown in Fig. 7.1 has been subjected (in theory) to the standard beam and torsion loads, i.e. 1500 lbf applied along the front seat cross-member, and a torque of 3000 lb ft applied at the centre-line of the front wheels. The resulting beam deflections and twist distributions are shown in Fig. 7.9. Fig. 7.10 shows the force and bending-moment diagrams for the side frames.

Analysis

A large volume of data can be obtained from the computer. One of the major problems lies in its analysis and useful application.

The deflections and aperture distortions can be plotted automatically and hence present little difficulty. They can be analysed on the basis of previous experience and any serious weaknesses in the body usually become immediately apparent so that modifications can be made.

The stresses present some difficulty. Firstly they are only really relevant if the load configuration is realistic. Secondly, they only apply to uniform sections, the programme taking no account of the stress-concentrations round joints. Critical areas of stress can be looked at in more detail by isolating them and making the idealized model more realistic. The loads calculated for the complete structure are applied to the boundary of each area.

However, the stresses obtained in beam and torsion tests are not very high. In practice they do not occur at a sufficiently high frequency to cause fatigue failures. But safety-test requirements (e.g. seat-belt anchorages) do produce stresses up to yield point and these usually require further investigation.

It is impossible to optimize the body by the usual methods because of the complexity of the structure and the number of sections with different properties: it has to be done by inspection. Either minimum uniform stress or minimum member-distortion can be used as an optimization criterion. Distortion is usually the critical phenomena, being related to noise and vibration, windscreen retention, door fit and sealing. When this is satisfactory, the stresses are usually low except locally round engine and suspension mounting points.

Taking the moments and forces as a criterion, many members appear to be unnecessary. Such members can be looked at individually to see if they are required for some other local loading condition. The ones that are still found unnecessary can be looked at in terms of their weight and construction, to see whether it is worth while modifying or eliminating them.

As a further stage in optimization, the strain energy of individual members can be calculated and compared with the total strain energy of the body. A high value indicates that the member concerned has a significant influence on the body stiffness. The converse does not necessarily hold, since members with low strain energy can still carry high loads. Using this technique, (**6**) it is also possible to estimate the effect of small changes in properties without re-running the complete programme.

It can be misleading to carry out stress optimization since many members which are highly stressed may be carrying little load. Hence they could be left out without significantly affecting other members. To do this, first eliminate non-essential load-carrying members, as previously suggested.

CONCLUSIONS

The method of structural analysis along the lines suggested has proved satisfactory. Its usefulness increases with

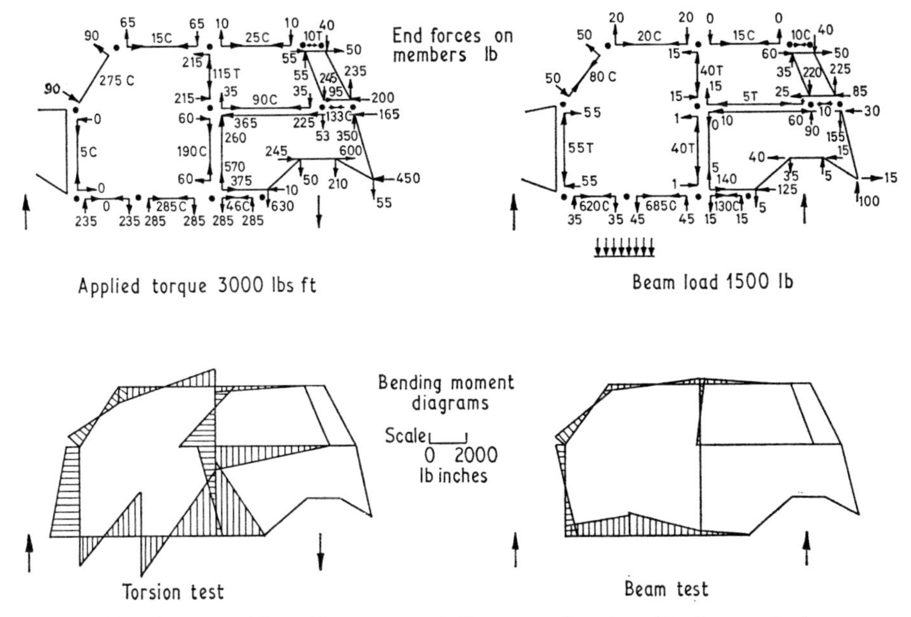

Fig. 7.10. Force and bending moment diagrams for the side frames in beam and torsion tests

experience as does its accuracy. A more quantitative knowledge is required of the behaviour of joints, the beam properties of open sections and the effectiveness of curved panels, before better consistency can be obtained.

When satisfactory procedures for idealization have been laid down, distributed-inertia loads must be investigated so that member-stresses will bear some relation to those occurring in practice.

True optimization of a body-shell is impossible. However, by trial and error it should be possible eventually to build a standardized inner structure of optimum strength, around which the stylist designs his skin panels. Each member would have predetermined properties, depending on the wheel-base and type of vehicle.

APPENDIX 7.1

REFERENCES

(1) Zienkiewicz, O. C. 'The finite element method in structural and continuum mechanics', McGraw-Hill.

(2) de Veubeke, F. 'Matrix methods of structural analysis', Pergamon Press.

(3) McKenna, E. R. 'Computer evaluation of automobile body structure', A.S.B.E. Annual Technical Convention, 1961.

(4) Allwood, R. J. and Norville, C. C. 'The analysis by computer of a motor car underbody structure', Proc. Inst. Mech. Engrs., Auto Div., 1965–66, Vol. 180 Part 2A, p. 207.

(5) Wardill, G. A. 'Computerized body design', Automotive Design Engineering, 1967 (Oct.).

(6) Dirschmid, W. 'A method of designing a vehicle body for optimum overall stiffness', A.T.Z., Vol. 71, no. 1 1969 (Jan.).

Paper 8

SMALL COMPUTER PROCEDURES AS TOOLS FOR STRUCTURAL DESIGNERS

G. A. Wardill*

The system of simplified computer procedures described is intended as a basic structural design tool for designers and draughtsmen. An example is given of how to estimate the beam-mode deflection of a production vehicle body. The degree of accuracy obtained is shown by comparison with rig test results. Data preparation and processing times are also discussed.

INTRODUCTION

THE INCREASING USE OF computers for structural design of vehicles is evident from published examples. These show that, given enough time for adequate preparation of data, very good results may be obtained.

Another paper delivered at this symposium shows some of the methods which have been evolved for the analysis of complete car bodies. It is evident that there is a place for this type of analysis within the industry—especially for long-term research into future body forms, where there are no precedents to give design confidence. However, as many engineers have found from bitter experience, there are still many problems to be overcome in the application of computers to assist in the design of current vehicles. There is an evident need for computer procedures which are easy to use and foolproof enough for draughtsmen and designers to use themselves. Such procedures would eliminate the delay caused by communication with structure specialists.

The aim of this paper is to discuss a possible system of computer procedures for the design office. Although examples of such a system in use are included, they are not intended to be a formal exposition of results obtained. The structural theory and programming aspects will not be discussed in detail here for both these aspects are covered by relevant A.S.A.E. Reports. Emphasis is placed rather on the needs of the drawing office and on describing the manner in which these needs have been met. Inevitably, the key is simplification of procedures. This must be so if a workable system is to be provided for non-specialist engineers, i.e. those who are not full-time structures analysts with computing experience.

When such a system is discussed, inevitably the field widens from the mere analysis of an automobile body. What, in effect, one is trying to achieve is to provide every designer with a set of graduated programmes, giving him virtually a 'tool kit'. These computerized 'tools' can be used in exactly the same way as, say, drawing instruments or a drawing board; they should become indispensable adjuncts to the professional career of the designer.

The proper use of any tool implies a period of formal training in its use. There is nothing magical about the computer or the programmes it uses. If they are used badly then poor work will result.

SYSTEM DESCRIPTION

Fig. 8.1 shows one complete system for automobile structure analysis. The analysis increases in complexity as the design progresses. This paper is only concerned with the first two stages; the analysis of plane frameworks, loaded in their own plane (side frame programme) and those loaded normal to their plane (grillage programme). Examples are probably the best way of explaining the part of the system examined. They also serve to show that these programmes can give realistic results for the deflections of a complete production car body.

Before such examples are considered, it will be helpful to describe the process of idealization and data preparation. This can be summarized in generalized terms as follows:

(1) Centre-lines are determined, corresponding approximately to the neutral axes of the various beams involved.

The MS. of this paper was accepted for publication on 3rd December 1970.
* *Cranfield Institute of Technology, Cranfield, Beds.*

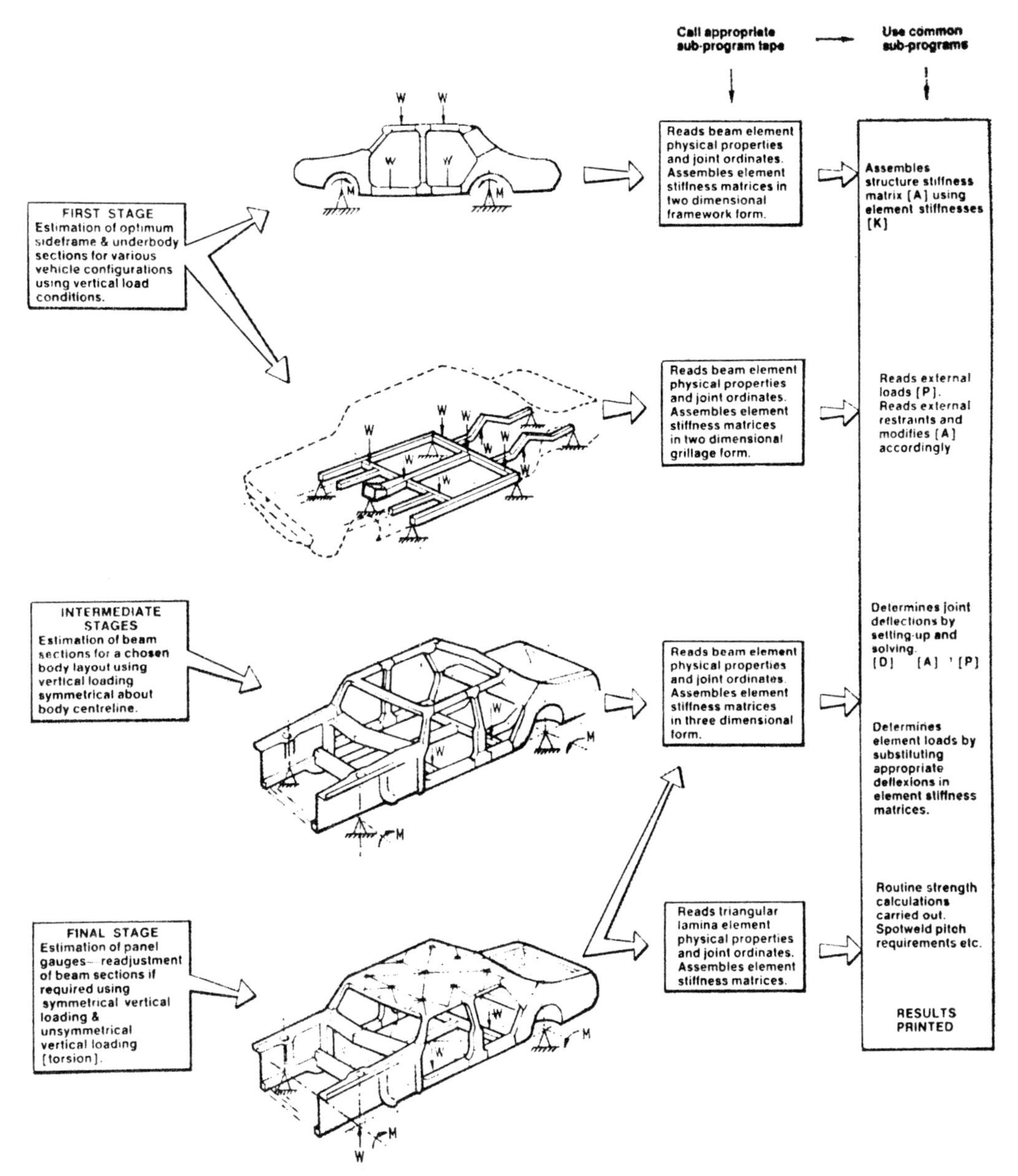

Fig. 8.1. A complete system for automobile structure analysis

(2) Nodal points are chosen at the areas of intersection of the beams, at points where section-changes occur or at points where external loads are applied.

(3) Gussets are chosen according to experience but a good start is the area around a joint covered by reinforcing material. Centre lines of connected beams need not intersect, the only requirement being that they run into a common gusset.

(4) Beam data are prepared describing the nodes, the cross-sectional areas, second moments of area, Young's moduli of the material, gusset dimensions, all relative to the local axes of the beam.

(5) Load and restraint data are prepared. These describe the magnitudes and directions of the applied loads and the required constraints on the movements of the nodes. These data are in terms of a set of reference or global axes.

The small amount of data required for these simple idealizations ensures that no costly delays are incurred during the preparation. In the general analysis of a complete body, data preparation time can be considerable.

EXAMPLES

During the early layout stages of the design, considerable time-saving can be achieved if certain body members under consideration can be shown to meet known stiffness and strength requirements. This type of information is

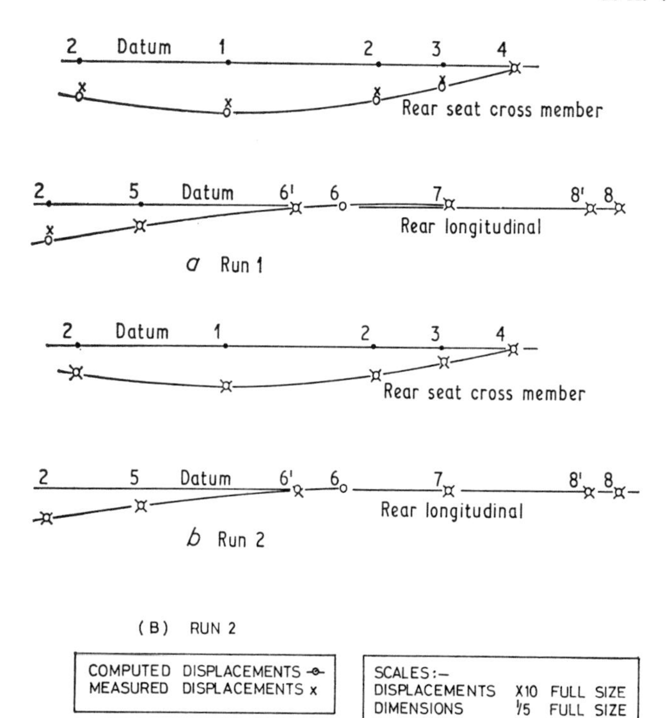

Fig. 8.2. These plotted displacements show that computed and test figures agree within possible manufacturing limits

particularly useful about such items as seat cross-members. Manufacturing or clearance requirements may, for example, result in a member with severely reduced cross-sectional area in some places. Adequate clearance for the propeller shaft may dictate such a reduction.

If a rapid analysis shows that the displacements and stresses for the member are comparable with previous satisfactory designs, then production planning can be undertaken with confidence at an early stage. This type of work is often carried out by specialist structures analysts. The necessary communication between the people on the drawing board and the analyst inevitably takes up valuable time. Where possible, direct evaluation of the structure by the designer, using the small-programme system, is an obvious advantage.

Rear floor

Work undertaken by the A.S.A.E. has shown that some areas, such as the floor assembly, can be examined in simplified form and in isolation from the rest of the vehicle. The first example quoted shows the results of examining the rear floor area of a production vehicle. The plotted displacement figures of Fig. 8.2 show that computed and test figures agree to within possible manufacturing tolerances. This being so, any attempt to increase the accuracy of the computation would serve no useful purpose.

The idealization shown in Fig. 8.3 is about as simple as it is possible to get, whilst still describing the

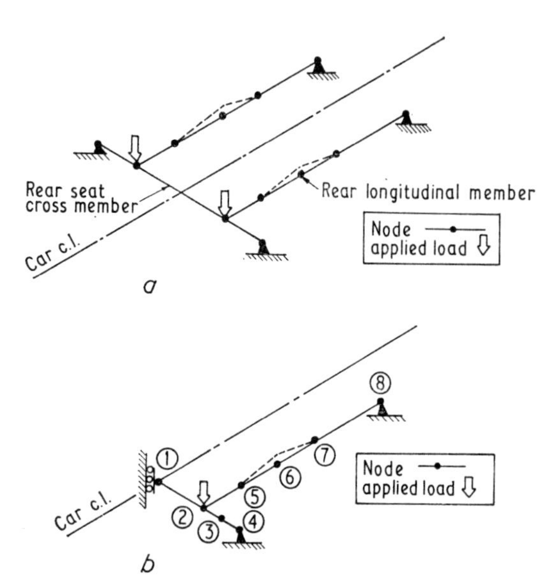

Fig. 8.3. (*a*) The schematic of the actual structure, and (*b*) the simplest idealization which adequately describes it

structure adequately. For example, three beams of constant sections are required to describe one half of the rear-seat cross-member. The degree of simplification is shown by the way in which the hump in the rear longitudinal members has been straightened out. This enables a grillage type of analysis to be used.

The calculation times achieved by using the automated procedures described in this paper can be seen in Table 8.1. It is interesting to note that while all the arithmetic is taken care of by the computer, human decisions are still by far the most important factor. For example, a choice still has to be made of the type of certain joints. Fig. 8.4 shows the joint at Node 4 of the floor idealization. In this case, complete freedom of rotation (by sagging) of the cross-member is allowed by the method of manufacture. Manufacturing requirements prevent the transfer of

Table 8.1. Automatic calculation times

DATA PREPARATION AND PROCESSING TIMES

The times shown are for the initial setting-up of programme data. It should be noted that subsequent computer runs for optimization purposes usually require only a few minutes of preparation (alteration) time.

(A) SIDE-FRAME PROGRAMME

The idealization uses seven nodes and eight beam elements (see Fig. 8.10). This involves the calculating of five section properties.

TIMES	
Section property calculations using special programme and teletype console.	$2\frac{1}{2}$ man-hours.
Node position calculations	$\frac{1}{2}$ man-hours.
Initial data tape punching (off line)	2 man-hours.
Main frame computing time (Univac 1108)	6 s

(B) UNDER-BODY PROGRAMME

The idealization uses eight nodes and seven beams. Data preparation and running times are similar to programme (A).

Fig. 8.4. The actual joint at Node 4 of the floor idealization shown in Fig. 8.3

Fig. 8.5. A direct attachment of the lower flange at Node 2 in Fig. 8.3, in which sagging rotation is restrained

CRANFIELD COMPUTING CENTRE
(Dept. of Mathematics)
PUNCHING INSTRUCTIONS
NAME
DATE.
Page 1 of 2
Application
DATA FORMAT COIL (SIDEFRAME)
TITLE LINE-UP TO 80 CHARACTERS
IIL
M N
Y
P
A
YT
YB
TT
PTN
X1
Y1
X2
Y2
X3
Y3
X4
Y4
1
2
IIL
REPEAT FORMAT OF ABOVE TWO LINES
ONE PAIR OF LINES PER BEAM
FOR IIL BEAMS
FOR EXAMPLE SEE FOLLOWING SHEET

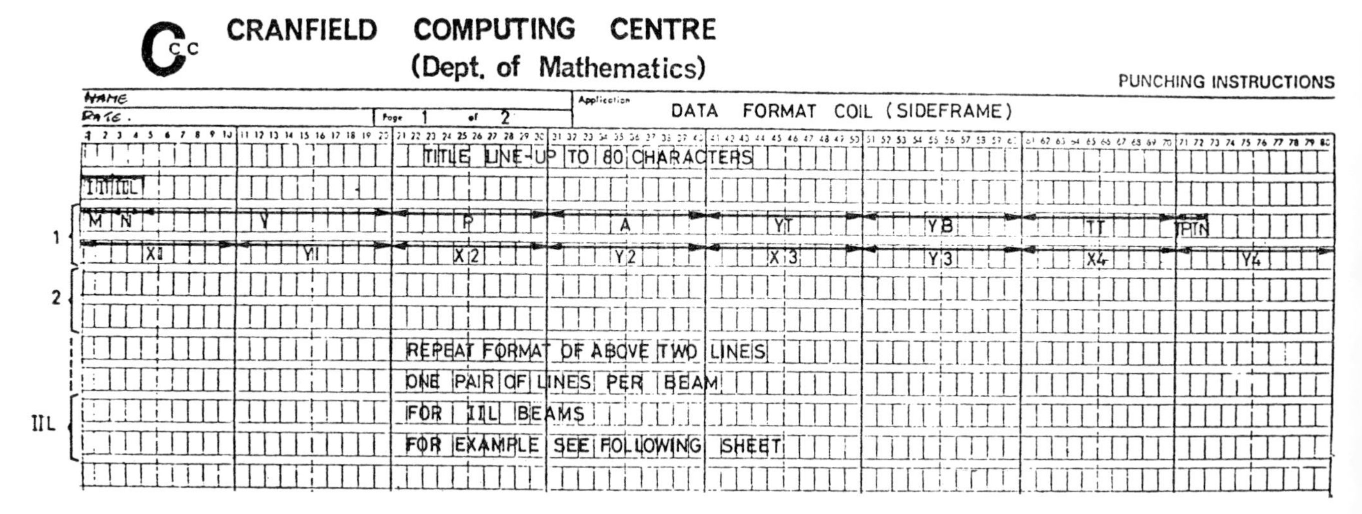

Fig. 8.6. Punched card layout for computer analysis of side frame

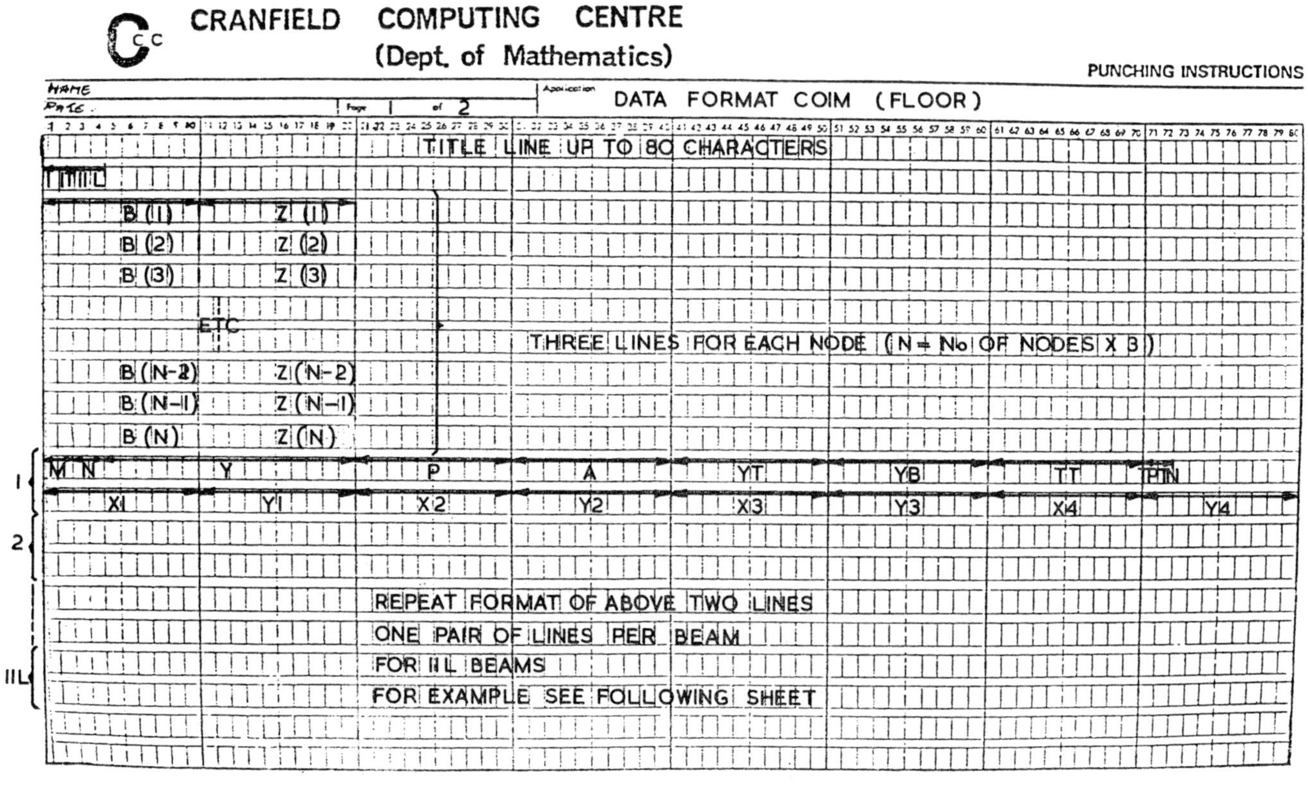

CRANFIELD COMPUTING CENTRE
(Dept. of Mathematics)
PUNCHING INSTRUCTIONS
NAME
DATE
Page of 2
Application
DATA FORMAT COIM (FLOOR)
TITLE LINE UP TO 80 CHARACTERS
TITLE
B (1)
Z (1)
B (2)
Z (2)
B (3)
Z (3)
ETC
THREE LINES FOR EACH NODE (N = No OF NODES X 3)
B (N-2)
Z (N-2)
B (N-1)
Z (N-1)
B (N)
Z (N)
M N
Y
P
A
YT
YB
TT
PTN
X1
Y1
X2
Y2
X3
Y3
X4
Y4
1
2
IIL
REPEAT FORMAT OF ABOVE TWO LINES
ONE PAIR OF LINES PER BEAM
FOR IIL BEAMS
FOR EXAMPLE SEE FOLLOWING SHEET

Fig. 8.7. Punched card layout for computer analysis of floor structure

bending moments to the supporting assembly by means of a direct attachment to the lower flange.

Fig. 8.5 shows a direct attachment of the lower flange at Node 2. In this case, sagging rotation is restrained.

Figs 8.6, 8.7 and Table 8.2 show the layout of the data required by the computer. The various joint conditions are easily described, by means of the symbols shown.

Side frame

The second example concerns a specialized programme for the analysis of the side frame alone. The results shown in Fig. 8.8 are typical of those which can be obtained. The agreement between computed and test results again is within production tolerances. The idealization diagram for this type of analysis is shown in Fig. 8.9. When instructed

Table 8.2. Symbols used in data sheets

IIT	= Number of nodes	(I_2 format)
IIL	= Number of beams	(I_2 format)
B(I)	= Load value	(F 10·0 format)
Z(I)	= Restraint	(F 10·0 format)
M	= Lowest node no.	(I_2 format)
N	= Highest node no.	(I_2 format)
Y	= Young's modulus	(F 16·0 format)
X	= Shear modulus G	(F 15·0 format)
P	= 2nd moment area about sagging axis	(F 10·0 format)
T	= Polar 2nd moment of area	(F 10·0 format)
YT	= Distance to point at which stress is required	(F 10·0 format)
Y_B	= Distance to point at which stress is required	(F 10·0 format)
X_1 =, Y_1 =, X_2 =, Y_2 =	Global ordinates of nodes	(F 10·0 format)
TT	= Wall thickness	(F 10·0 format)
IPIN	= O for rigid joint connection; I for flexible joint connection	(I_2 format)

to do so, the programme will automatically assume rigid body movements for sections other than the side frame. It will also assume rigid gussets at certain joints within the side frame, if told to do this. Data layout and quantity is very similar to that required for the floor-pan example, show in Fig. 8.7 and Table 8.2.

Work by the A.S.A.E. has shown that it is possible to combine the results of analyses of local areas in order to predict more general characteristics of the body. For example, the results of the two examples quoted can be combined, if necessary, to estimate the beam mode of deflection of the whole vehicle. Fig. 8.10 shows the idealization required for this procedure. The results, shown in Fig. 8.11, again lie within reasonable production tolerances. In other words, a very good idea of an important aspect of the flexibility of the vehicle can be obtained in a very short time.

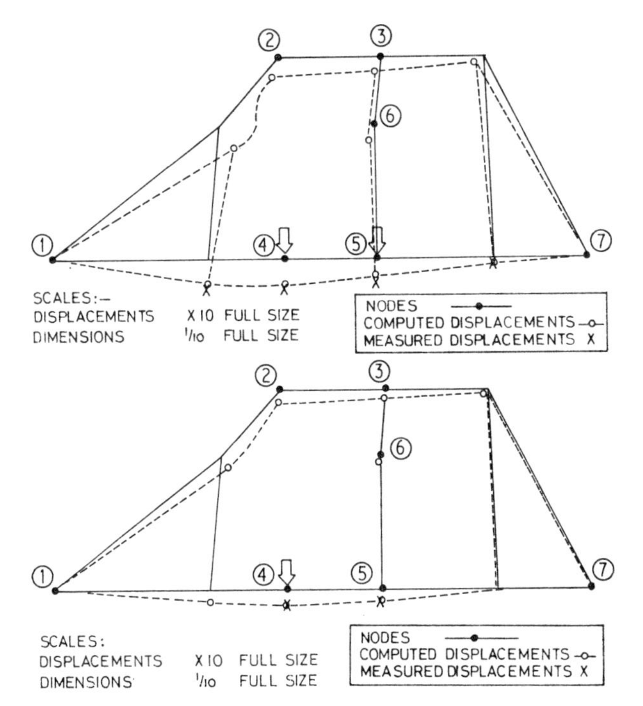

Fig. 8.8. Typical results of side-frame analysis: computed and test results again agree within manufacturing tolerances

Testing

For verification, all computed results have to be compared with test results. If production components are tested, extensive rig facilities must be provided. All experimental figures quoted in this paper are the result of A.S.A.E. tests. A programme of such tests is being undertaken with the aim of determining further simplified computer procedures. Fig. 8.12 shows the vehicle-body test rig. A production body in 'white' is used, i.e. minus windows, mechanical fittings, trim and paintwork. The test-rig shown here is of the type which provides support, leaving the body free to sag or distort in any of the lozenging modes, for example. The results quoted were obtained by loading the body at the appropriate stations and measuring resulting displacements by dial gauges, independently.

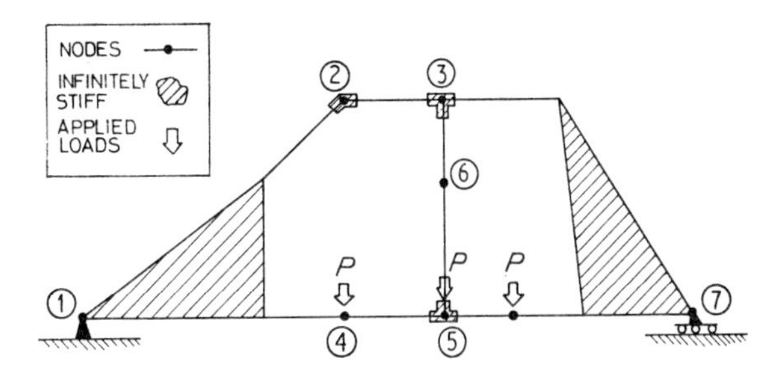

Fig. 8.9. How the side frame is idealized for computer analysis

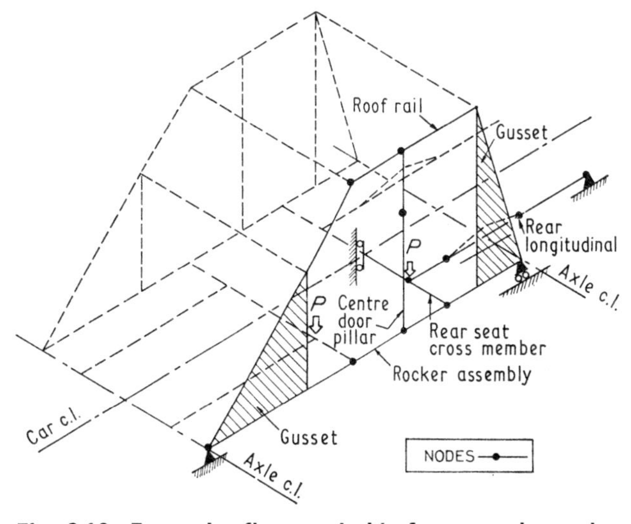

Fig. 8.10. From the floor and side-frame analyses described above, the beam mode of deflection for the whole vehicle can be computed with the aid of this idealization

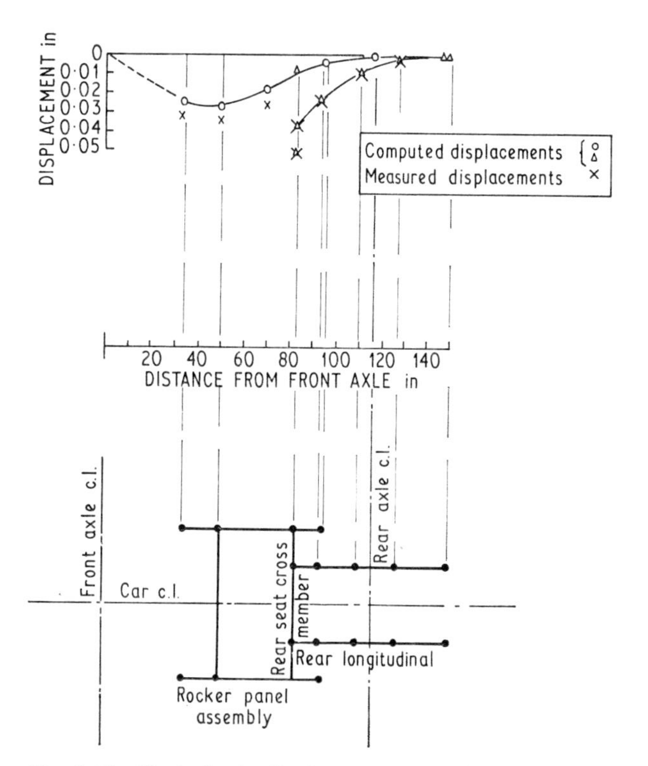

Fig. 8.11. Underbody displacement obtained from the analysis of Fig. 8.10

Load tests are also carried out on sub-assemblies which are mounted in the cathedral-type rig, shown in Fig. 8.13. Here a side-frame assembly is being subjected to side-loading of the type which might be experienced under side impact.

In addition to continuing the type of testing mentioned, work on simplified torsional analysis of bodies is planned.

Examples given so far have shown how certain procedures in the design of a new body can be categorized and included in a computer system. As we all know, structures designers constantly run into a host of minor strength and flexibility problems which are impossible to categorize but which nevertheless take up a large part of the design time.

The solving of these problems is greatly facilitated by a system of small programmes such as is described here. A brief mention of just one example will illustrate this point. The door frame above the waist line will often be reduced to minimum proportions by styling requirements. There is nevertheless an appreciable load applied by the existing sealing as shown in Fig. 8.14. Estimates of distortion and stresses in this area (and indeed of the whole door, if required) can be carried out very quickly by using, say, the underbody programme. The door is in fact an example of a very localized problem which can be dealt with quite separately from the rest of the body. Time taken for this

Fig. 8.12. The A.S.A.E. body test rig for determining the experimental data used in this paper

Fig. 8.13. A side frame being subjected to side-loads such as are experienced in side-on collisions

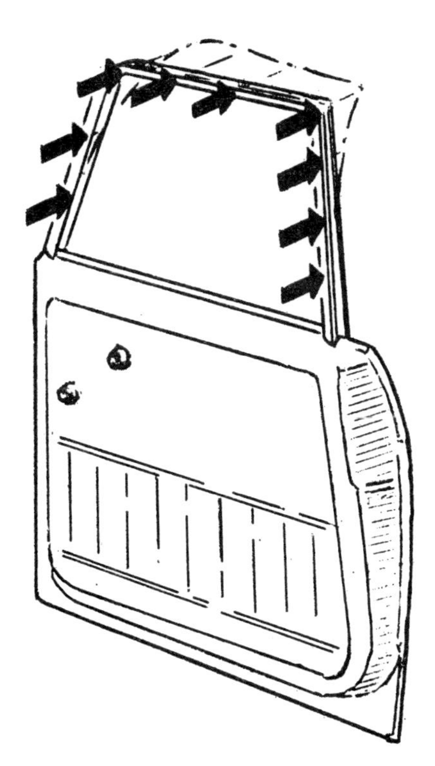

Fig. 8.14. While the door window-frame is reduced to a minimum by considerations of styling, it has to carry appreciable sealing loads

type of analysis can be as little as a few minutes (again assuming good communication with the computer).

This type of calculation, when done by hand (e.g. by strain energy techniques), could easily absorb a man-day. Many other examples of this type of small problem could be quoted, including those arising on items such as suspension components.

PROGRAMME MAKE-UP

The primary aim of this paper is to show practical applications. Nevertheless some description of the make-up of the programmes is desirable since they have been written specially to deal with these applications.

Basically, the programmes permit the operator to build a software model of the structure, by assuming an assemblage of rigid blocks and flexible elements, such as beams or plates. This concept, and the need to allow for the easy description of joints of various stiffnesses, are the main reasons for the writing of special programmes. Mathematically these assemblies can be described by means of matrices. In the programmes described, the matrices are built-up by means of the direct stiffness method.

Now a combination of mathematics and structural theory can result in some very powerful methods for the assembly and manipulation of the necessary matrices. The programming required is sometimes very complex. In fact many advanced structural analysis programmes, such as the ones under development at the A.S.A.E. and other centres for research on large complicated structures, take man-years to write. These programmes can only be properly understood and used by specialists.

The type of system described here has no need of such advanced procedures. Indeed it possibly is undesirable to use them for the following reasons:

(1) It has been shown that small localized areas of the body can be looked at separately. Considerable savings in data preparation time can be effected by simplified procedures for which only a small amount of computer core-store is required. Thus the automatic matrix formulation can be straightforward although somewhat inefficient from a storage point of view.

(2) Experience has shown that, if the matrix routines are kept simple, non-specialist engineers can quickly grasp the basis of the programmes and the structural theory behind them. By understanding the underlying principles thoroughly, engineers can increase the effectiveness of their work; mainly by learning to avoid certain basic errors in their idealizations. It has even been found that, after a two-week concentrated course, some people are able to modify the matrix routines, in order to fulfil some specialized need.

There are parts of the programmes which are inevitably complicated and difficult to understand; areas such as the mathematical solution routines, automatic substitution routines, etc. These need not be completely understood by the operator. Here it is only essential that he understands certain diagnostic routines. For example, certain types of structures could be impossible to deal with, even on a computer. The operator must know when the computer has 'given up the ghost' and is producing nonsense. In solving the matrix equations, certain checks are given to the operator in the form of a residual vector. With a little training he can quickly determine whether the answers are sensible or not.

An incidental but important advantage arises from the efficiency with which data is stored and can be retrieved. When data for any segment of the body is fed to the computer, a very simple routine can be used, which will compare it with previous data. If there are any differences, the computer can be made to question them and, given a satisfactory reply, the store can be updated.

Thus an up-to-date record can be kept. At any time it is possible to obtain, say, isometric layouts of centre lines for the body to date by calling on the store and obtaining a print-out via an on-line plotting device. It would be possible, even with present-day technology, to employ a data acquisition service linked to the system. A light-pen device could be used at design meetings. By calling up the appropriate simplified programme and up-to-date information, the effects of making certain changes to the structure could be seen almost instantly and decisions could be made whilst the meeting was still in session.

CONCLUSIONS

The main lesson to be learnt when reviewing past experience seems to be that research into theoretical and programming techniques is not enough on its own. If computer techniques are to be successfully introduced into the design process, just as much work must be put into formulating a complete system. This should take into account:

(*a*) The degree of accuracy required.

(*b*) The techniques required to simplify and subdivide the analysis, with a view to keeping data preparation time to a minimum.

(*c*) The simplification in programming required to enable the users to understand, and even modify, procedures.

In addition, there is a need to provide adequate background instruction for the users. Very often engineers are sent on a one-week general programming course and are then considered capable of making full use of computers. It is suggested that, in the structural field at least, efficient computer utilization will only be achieved if more comprehensive instruction is given on the background of appropriate simplified programmes.

In this paper I have tried to show some of the uses of a system of small programmes which satisfies at least in part the conditions listed above. The very essential background educational techniques have not been discussed but, it is hoped, at least the needs in this area have been highlighted.

Paper 9

THE STRUCTURAL DESIGN OF BUS BODIES

G. H. Tidbury*

The problems of designing bus and coach bodies are discussed. The matching of the body structure and chassis frame causes problems which have been largely overcome on an ad-hoc basis. Simple theoretical methods are suggested for estimating the stiffness in bending and torsion of the superstructure of composite buses. The formulae are checked against computer analyses and plastic models when possible. The proportion of load carried by the bodywork is estimated to find the stresses in critical members for any combined loading. Although further analysis and full-scale testing are required, preliminary design calculations for composite buses can be based on this method.

INTRODUCTION

THE STRUCTURAL design of buses has always received more attention in technical publications on the Continent than in Great Britain and it is of interest to speculate about the reasons for this.

First, the number of people using buses as their only means of transport is higher in Eastern Europe than in the West and this would account for the importance of bus design in those countries. Secondly, for several decades British roads have enjoyed a higher standard of surfacing than those of most other countries. Bad surfaces cause high torsional loading on vehicle structures and torsional loads, as will be shown, are carried more by the bodywork than the chassis. Therefore, unless torsional loads are small they must be carefully analysed by the designer of chassis-type vehicles.

When bodies are strong enough to resist high torsional loading, they will generally also carry the bending loads and this explains the popularity of the integral vehicle where roads are bad. It is notable that the 'Olympic' bus, produced in Britain specifically for operation in rough road territories, is an integral vehicle.

Another solution is to build a flexible body on a separate chassis and in some countries wooden bodies mounted on rugged chassis have a satisfactory life.

Because of the trend toward the integral bus on the Continent, several methods for stress analysis have been proposed for it. It is probable that a substantial part of the bus industry in Britain will continue to mount bodies on separate chassis and for this reason the attempt has been made to incorporate some of the simple concepts used by Continental designers for integral buses in the analysis of composite vehicles. These concepts are checked against computer analysis and model visualization, where possible.

The MS. of this paper was received at the Institution on 16th June 1972 and accepted for publication on 26th June 1972.

* *Senior Lecturer, Advanced School of Automobile Engineering, Cranfield Institute of Technology, Cranfield, Bedford.*

STRESS ANALYSIS

Methods proposed

The most sophisticated analysis of bus structures seen by the author was published by Brzoska (**1**)† in Polish in 1955. This work is mainly concerned with integral buses and uses the best techniques of aircraft fuselage analysis available at that time. Brzoska analyses structures of the forms shown in Figs 9.1, 9.2 and 9.3.

For the type shown in Fig. 9.1, it is assumed that the structure above the waist rail is so flexible, relative to the underbody which is reinforced by transverse tubes, that no load is taken by the roof. Fig. 9.2 illustrates a fully integral structure where only the end bulkheads are stiff in their own plane and where the torsion loads are applied in that plane. The third type analysed by Brzoska is an integral bus in which all the bulkheads are flexible in their own plane (Fig. 9.3).

All the structures analysed are open-ended tubes with no door openings. This has meant that Brzoska's paper has not been widely used in spite of the fact that his analysis is more detailed than those of other writers. (He includes the effects of the non-linear stress-distributions in shell structures.)

Hungarian workers, notably Michelberger (**2**), have designed successful buses utilizing an underfloor grill

† *References are given in Appendix 9.3.*

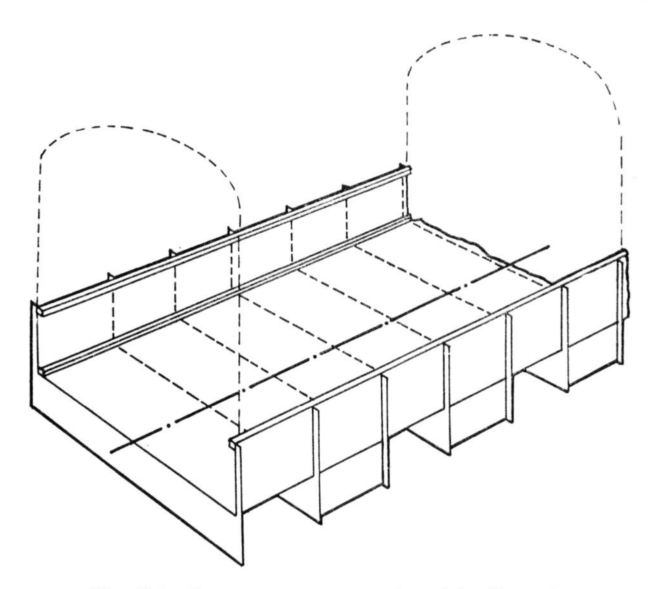

Fig. 9.1. Open structure analysed by Brzoska

attached to side walls with pin joints (Fig. 9.4). Again the assumption is made that no load is taken by the roof. The members of the floor grill are assumed to carry loads in torsion and bending and to be rigidly jointed in both these senses. The analysis has been extended (3) (4) to include the effects of large door openings where the loads are absorbed by the underfloor members instead of by the usual reinforcing ring round the door. Michelberger's analysis included vehicles with separate chassis (Fig. 9.5), but only in so far as the load-carrying structure of the body was assumed to be in the floor cross-members and the side-walls up to the waist-rail.

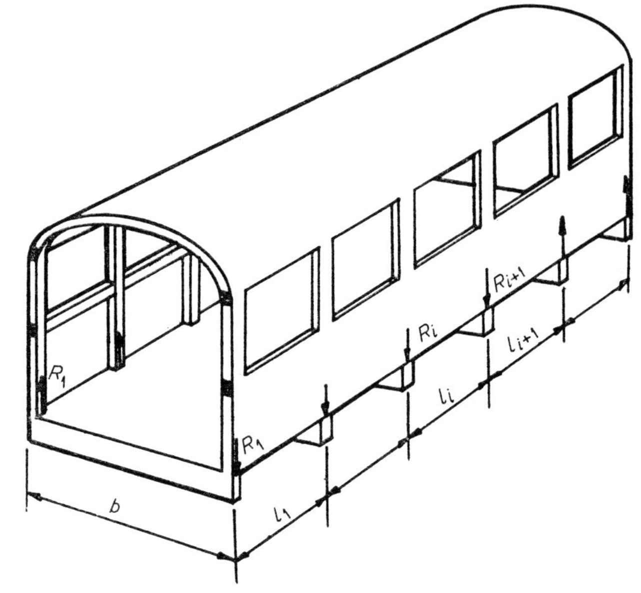

Fig. 9.3. Integral structure with intermediate rings

In a classical series of papers on commercial-vehicle stress-analysis, published in 1957, Erz (5) included a section on bus design. After describing alternative methods (Fig. 9.6) for underfloor structures that can be sufficiently stiff to relieve the superstructure of loads, he proposed a particularly simple method for estimating the critical design loads for an integral vehicle. This assumes that the side-walls carry the main loads in both bending and torsion and that the structure below the waist-rail is infinitely stiff, compared with the beam elements surrounding the windows and doors.

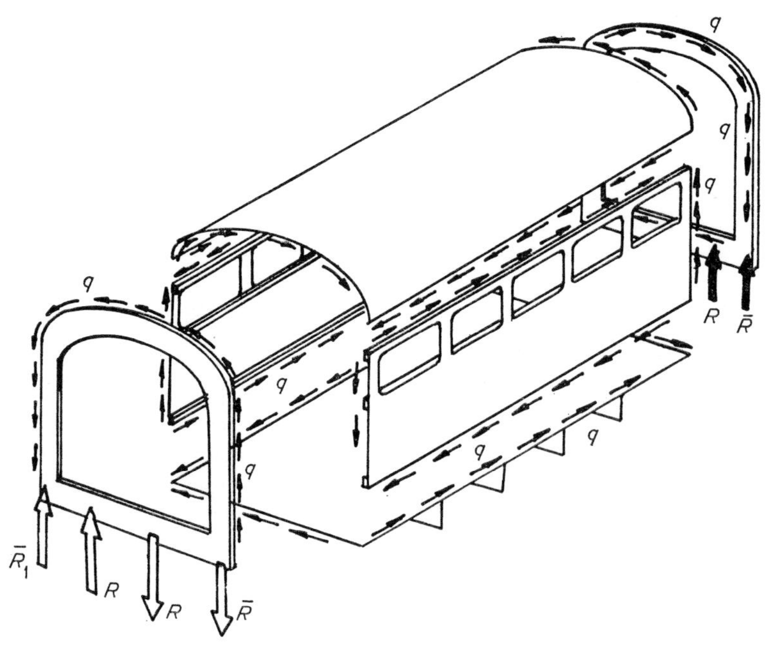

Fig. 9.2. Integral structure with end bulkheads only

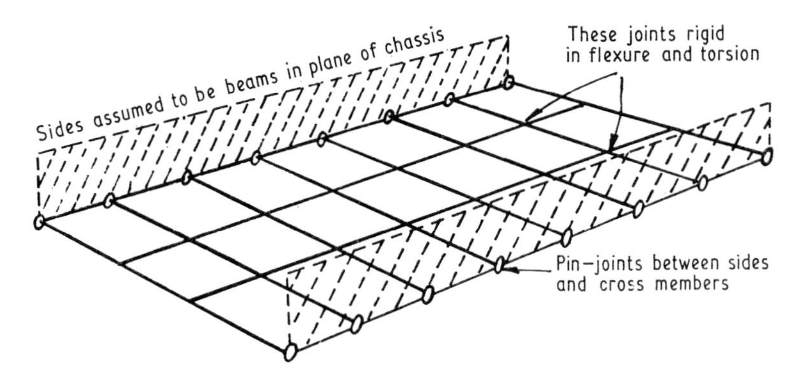

Fig. 9.4. An underfloor grillage is attached to the side-wall by pin joints in Michelberger's bus

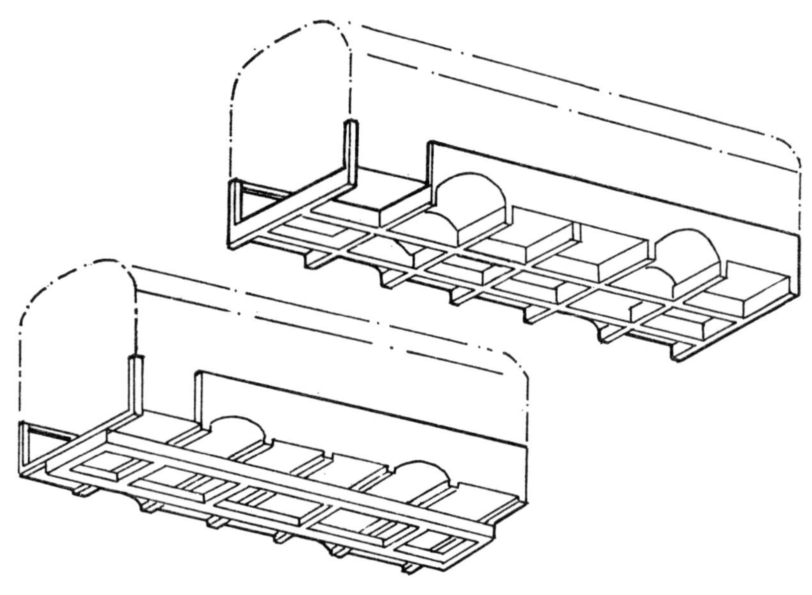

Fig. 9.5. Michelberger considered vehicles with a separate chassis but assumed load-carrying members only up to the waist-rail

In bending, the door-openings and window-pillars are the critical areas and approximate formulae are given in Fig. 9.7 for the maximum bending moment in the window-pillars due to bending, and in the cant-rail and in door sills as a result of the shear force. In Fig. 9.8 the method of analysis for torsion is shown. It will be noted that the structure is treated as a thin-wall tube with its axis across the vehicle. The transfer of shear from the cant-rail to the waist-rail again puts the window-pillars in bending.

Schemes for the complete analysis of the various types of bus structure discussed so far are given by Pawlowski (**6**) and new ways of incorporating torsion boxes are also included (Fig. 9.9). When dealing with the problem of transfering shear from the cant-rail to the waist-rail by bending in the window-pillars, Pawlowski suggests the use of the two extreme values of bending moment:

(*a*) where there is a pin joint at the cant-rail or the cant-rail is flexible in bending, and

(*b*) when all joints are rigid and the cant-rail beams are assumed to be completely rigid.

In case (*a*) a bending moment arm equal to the total height of the pillar is obtained while in case (*b*) the maximum arm is only half that height. Erz suggested that an empirical factor (*c*) of two-thirds be used for computing this bending moment. On the other hand the bending moment at the base of a portal with equal stiffness in all members and legs carrying a similar load in shear (*d*) suggests a factor of 4/7, approximately midway between the alternative suggestions (Fig. 9.10).

It must not be assumed from the references quoted that no stress analyses have been carried out in Britain but in most cases these have not been published. Further, the references quoted have all been to simplified analyses without computers (although the floor-grillage idealization of (**2**), (**3**) and (**4**) lends itself to a standard computer solution).

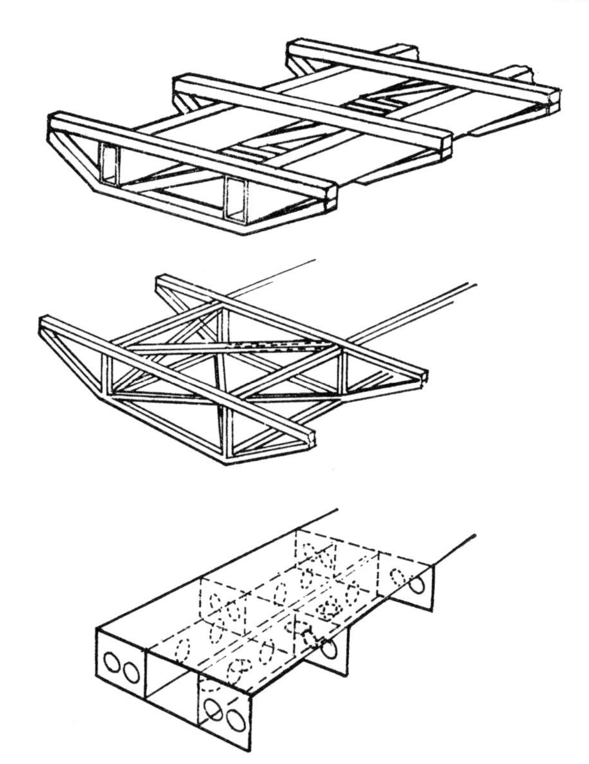

Fig. 9.6. These alternative underfloor structures are stiff enough to relieve the superstructure of loads

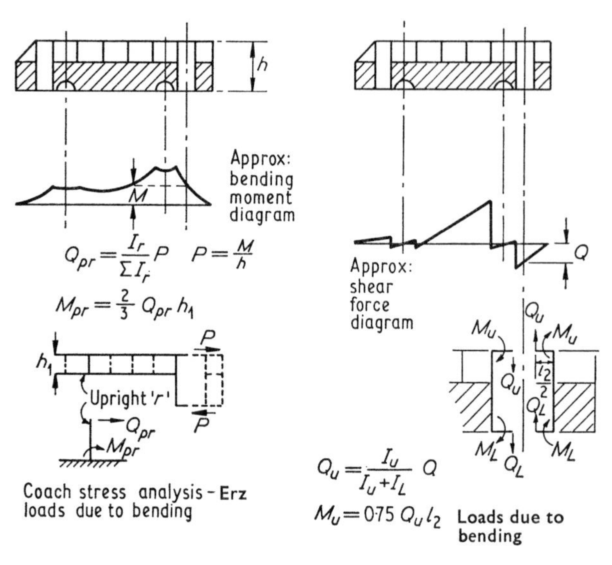

Fig. 9.7. Maximum bending moment in window-pillars, and shear force in door sills

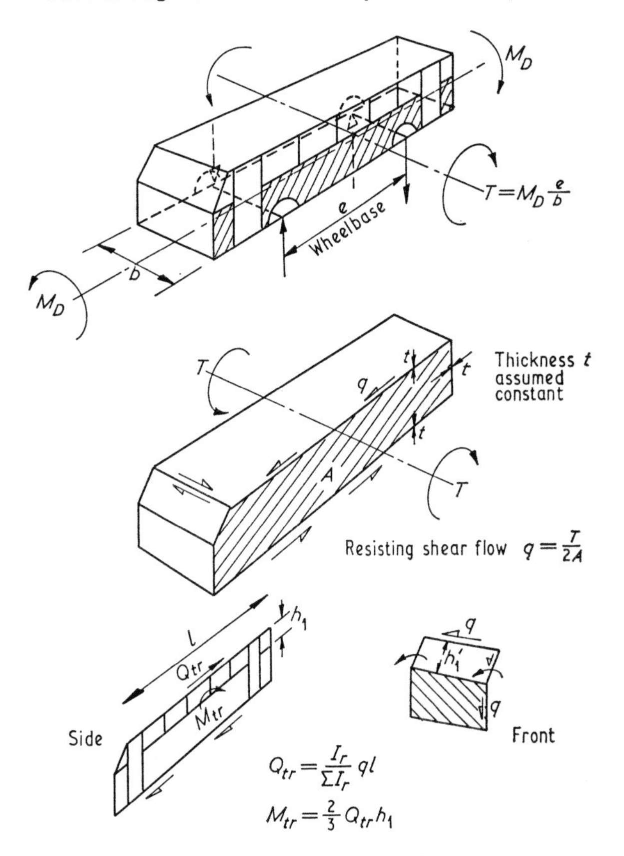

Fig. 9.8. Torsion analysis assumes the coach to be a thin-walled tube with a transverse axis

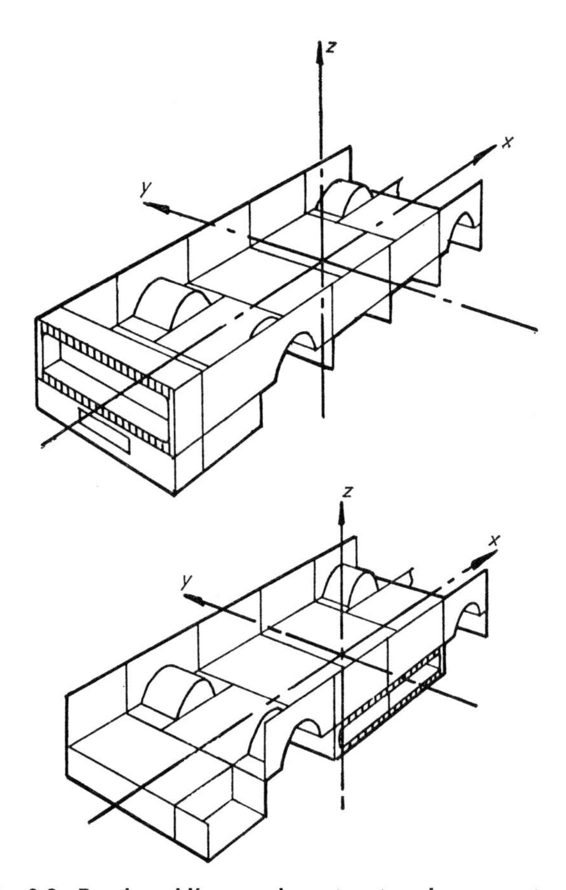

Fig. 9.9. Pawlowski's open bus structure incorporates torsion boxes in a new way

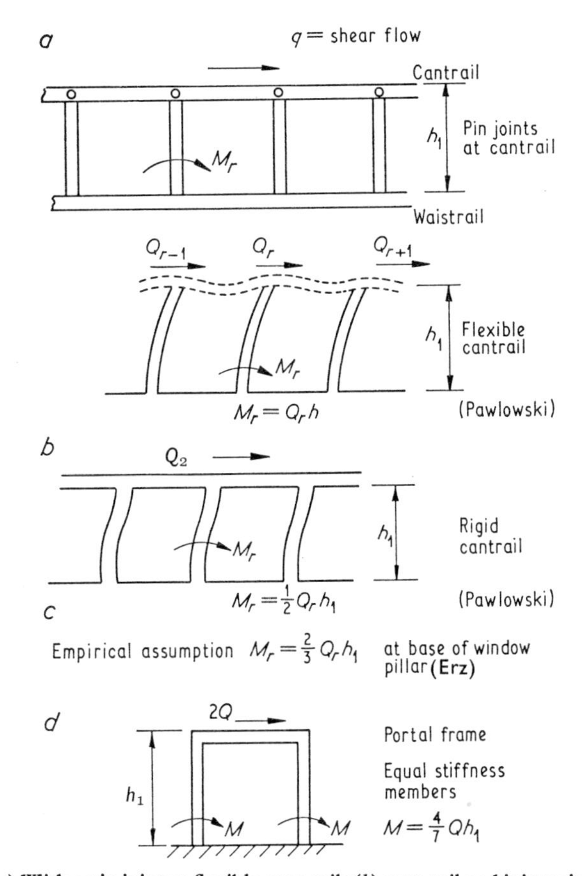

(*a*) With a pin joint or flexible cant-rail; (*b*) cant-rail and joints rigid; (*c*) an empirical factor; (*d*) the portal frame assumption gives a factor half-way between the others.

Fig. 9.10. Bending moments in window-pillars

Other workers, notably Alfredson (**7**), have published complete computer analyses of integral buses using finite element methods. An early attempt at a similar analysis of a fictitious bus structure was carried out at the A.S.A.E. for a student thesis (**8**).

Finite-element analysis can be made as complicated as the computer will allow but must always be used as a check against a proposed design; when used for a complete three-dimensional analysis the method gives the designer little guidance on how to improve a structure. However, as will be seen, simple redundant frameworks can be analysed by the computer programmes for side-frames and grillages given in (**9**), and the results incorporated in non-computer analyses to give a useful picture of how the main loads are carried in a mixed structure.

Bending

Vehicles with no doors between the axles

The bending loads in a composite vehicle will be shared between the chassis frame and the bodywork in proportion to the stiffness of the two structures. In practice this means the stiffness of the side-frame of the body and the

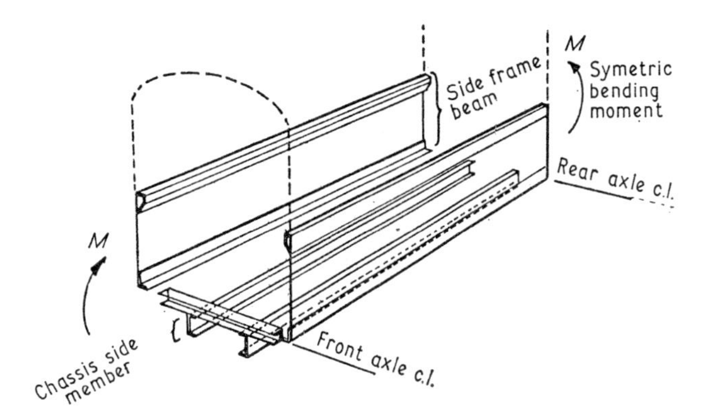

Fig. 9.11. With symmetrical bodywork the stiffness of the chassis side-members and body side-frames are comparable

side-members of the chassis. If the bodywork is symmetrical, as is nearly the case for touring coaches, the beam formed by the body-side between the window sill and the floor can be compared with the chassis side-members (Fig. 9.11).

For an all-steel body the ratio of stiffnesses will be in the order of 4·2 to 1, for an aluminium body 1·4 to 1 and for a wood-framed body it may be as low as 0·000 64 to 1, assuming the cladding only carries shear and the bending stiffness depends only on the wooden members. These simple comparisons depend on two assumptions: first, that there are no door-openings or wheel-arch cut-outs over the length of the vehicle considered (normally between the axles); and secondly that the floor cross-members and the attached pillars and roof-bows form bulkheads that are infinitely stiff in shear.

The first assumption is normally true for touring coaches but the overhang loads form a separate and important loading case for the side-frame over the rear wheel-arch where its depth is reduced. The second assumption is not strictly true, more of the bending load being carried by the chassis frame than would be indicated by the ratios quoted.

It is clear from the sample bending moment diagram in Fig. 9.7*a* that a door-opening in front of thc front-axlc will be in the area of low bending moment and shear (particularly for a rear-engined vehicle) and no difficulty would be expected in providing an adequate structure round the door-opening to carry the small loads.

Vehicles with a door between the axles on one side

In buses for standing passengers the door-openings on one side need to be larger so that an asymmetric structure results. A particular case investigated was the Ha'argaz bus, shown in Fig. 9.12. This vehicle was the subject of an early attempt by Hochberg (**10**) at the type of calculation described here. It is mounted on a chassis with the engine in the middle and is typical of the all-metal bodywork used for both city and rough roads.

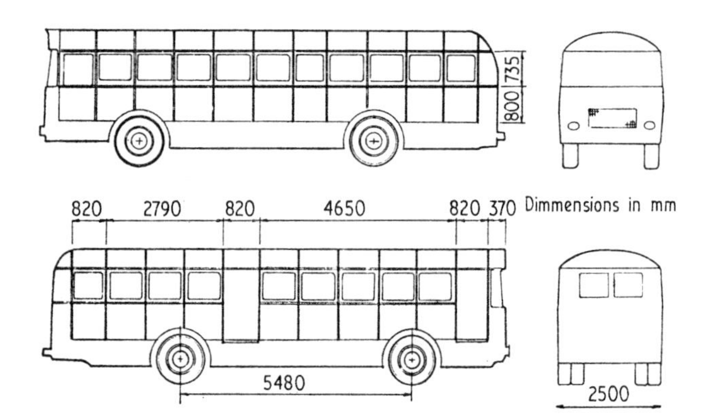

Fig. 9.12. A typical asymmetrical bus with large doors: the first step in analysis was to compare the stiffness in bending of the two different body-sides

Since the bending stiffness of the asymmetric body and the chassis frame must be compared to assess the proportion of the load carried by each, the first approach was to compare the stiffness of the two body-sides under pure bending. The Erz assumption (Fig. 9.7*a*) was used for the side with a door-opening and it was found that the stiffness of the window-pillars in bending, while transferring the compression load in the cant-rail, was comparable with that of the side-wall beam between the floor and the waist-rail.

This rather surprising result was confirmed by tests on polystyrene models (Fig. 9.13). The photographs are double exposures, taken before and after the load was applied. In the models no attempt was made to simulate the bending stiffness of the side-wall between the floor and the waist-rail, but the window-pillars, cant-rail and door-sill are in proportion to the respective second moments of area in the actual structure.

In practice the bending moment on a vehicle is always generated by vertical loads and the pure moment case considered above does not arise. Deflection of each side was therefore compared under a single load midway between the axles. The polystyrene models now showed about three times the deflection (Fig. 9.14).

The bending stiffness of the side-walls between the floor and the waist-rail has been correctly represented in Fig. 9.14 to the same scale as the pillars in Fig. 9.13. A calculation based on the assumption made by Erz (Fig. 9.7*b*) on the way the shear force is carried over the door-opening gave a ratio of flexibilities of the two sides as 3·15 to 1.

Since the deflection due to the window-pillars for a bending load on the structure, calculated as in Fig. 9.7*a*, was small, it can be assumed that, for this particular

Fig. 9.13. Polystyrene model of the bus in Fig. 9.12. Double exposures (before and after loading) confirmed that the stiffness of window-pillars in bending was comparable to the whole side-wall up to the waist-rail. A pure moment was applied

vehicle, stiffness in bending of the side-frame includes some contribution from the cant-rail, giving an increased bending modulus. This increase is shown in Fig. 9.15 as an arbitrary factor but it is based on the computer result for a side-frame without a door, shown in Fig. 9.16. The computed deflection indicates that the ring round the door-opening deflects in a complicated way. The 50 per cent increase in the bending moment of the cant-rail and door-sill, suggested by Erz in Fig. 9.7*b*, does allow approximately for the deflection of this ring.

The two sides with different flexibilities are joined by a series of rings and the flexibility in shear of these rings will affect the overall flexibility of the body. If the sides of the rings are very stiff, the shear will be carried by the

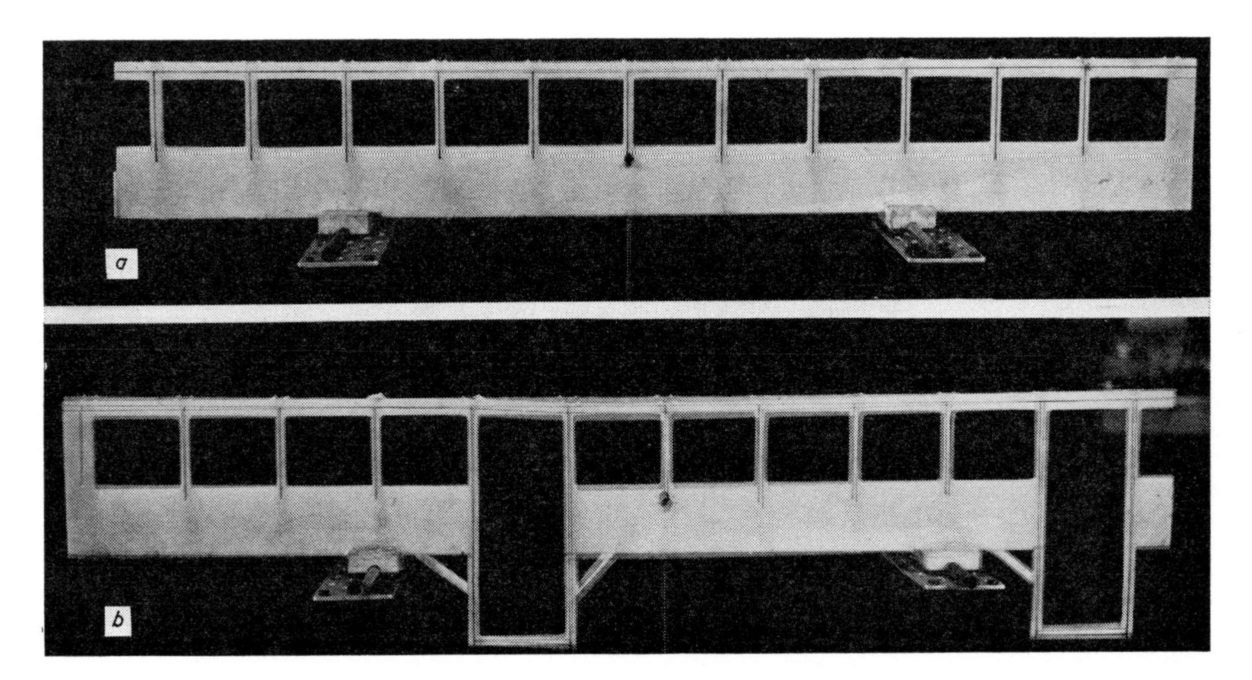

Fig. 9.14. Model side frames with central load applied: all beam stiffnesses are to scale

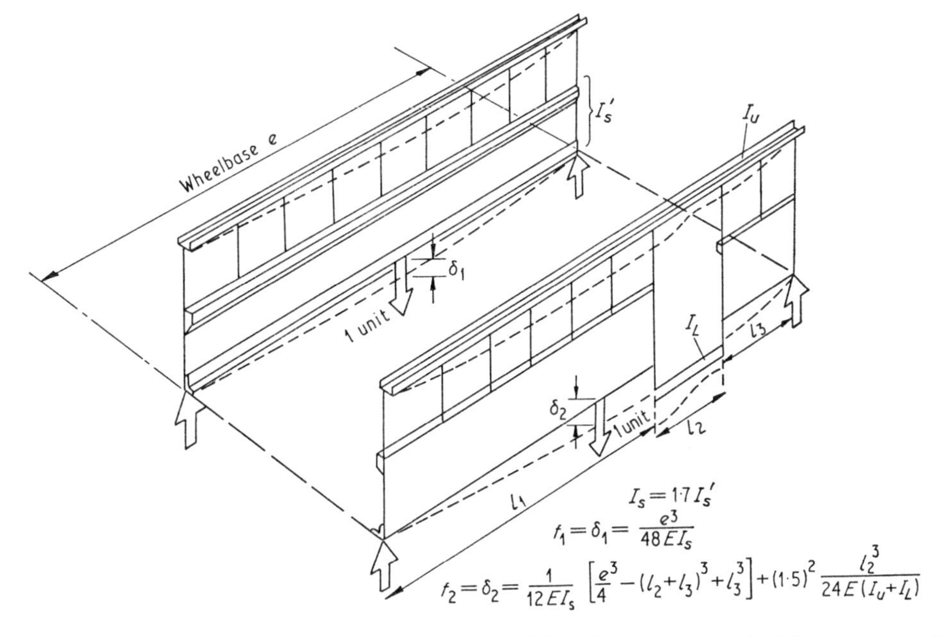

Fig. 9.15. The bending flexibility of the two sides of an asymmetrical bus compared

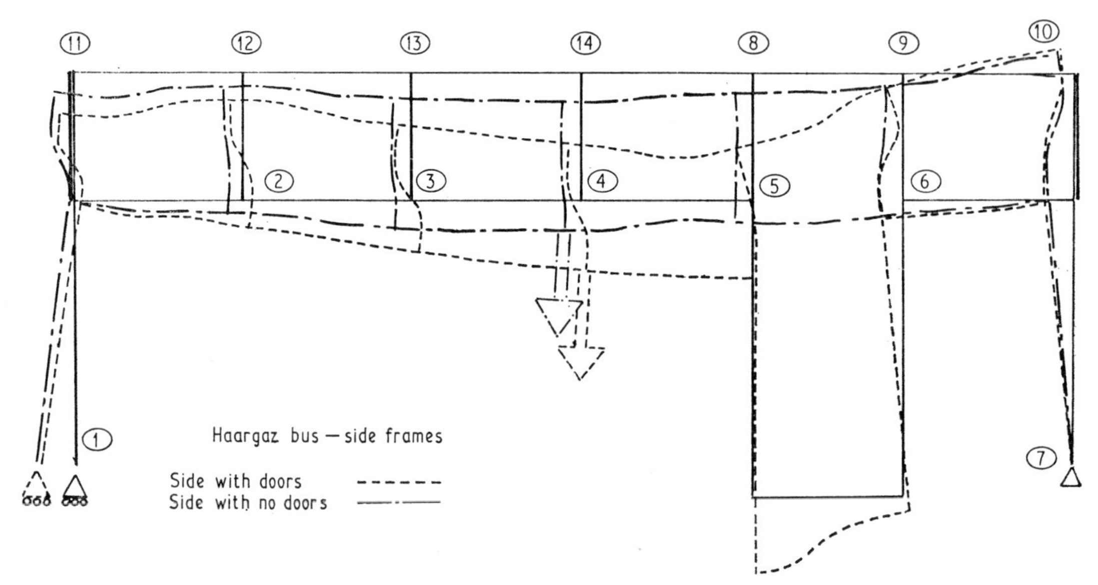

Fig. 9.16. Computed deflections for side-frames under central load

roof-bows and floor cross-members, in the same way as the cant-rail and door-sill carry the shear over the door-opening in the side of the frame.

The effect of one such ring at the centre of the wheel-base can be calculated by the formula given in Fig. 9.17. The flexibility (f_3) of this ring can vary widely with the stiffness of the sides. The effect of adding the other rings can only be found by a grillage analysis.

This has been carried out on the computer and the result compared with the result from the polystyrene model as shown in Fig. 9.18. The computer gave the ratio of the deflections of the two sides as 1·64 to 1, and the model 1·8 to 1. To obtain the former it is necessary to add the stiffnesses of all the rings in the structure and further increase the stiffness ($1/f_3$) by a factor of 1·75. In fact, the overall flexibility of the combined structure is not very sensitive to the value of f_3, and this analysis indicates that a good approximation can be obtained by taking f_3 as the inverse of the sum of the stiffnesses of all the roof-bows and floor-members which transfer shear between the two sides, as indicated in the final formula in Fig. 9.17.

Application of these formulae to the Ha'argaz bus gives a ratio of body to chassis stiffnesses of 7·4 to 1 so that 0·88 of the total bending load will be carried by the body. Because of the effect of the rings joining the two sides, the bending load carried by the body will be much more evenly shared than would be indicated by their flexibilities. The ratio of the loads carried by the sides is 1·32 instead of 3·15 to 1.

Torsion

The concept in Fig. 9.8 leads to a very simple analysis for the torsional stiffness of the body-shell of a bus. The floor and roof are assumed to be infinitely stiff in shear and all deflection takes place in the side-frames and the front and rear ends. The assumption is also made that the panelled structure below the waist-rail is stiff in shear compared with the window-pillars.

By making three of the four vertical frames infinitely stiff in turn, the overall flexibility of the structure in torsion can be found when the window-pillars in one frame (side- or end-frame) are deflected in bending. The formula is:

$$\frac{\theta}{M_D} = \frac{e^2 Z^2 h_1{}^3}{27 A^2 b^2 E \sum I_r} = f_i$$

The notation is defined in Fig. 9.8 and $\sum I_r$ is the sum of the second moments of area of the window-pillars in the flexible frame.

The overall body-shell flexibility is then:

$$f = \sum_{i=1}^{4} f_i$$

The formula allows for asymmetrical side-frames above the waist-rail, heavy door-pillars, panels incorporating route information, etc., on one side only. It does not, however, allow for the increased flexibility when the door-opening extends below the waistline and further work would be required to find the correction for this effect.

Based on the above formula, the torsional stiffness of the bus body analysed was 5000 times as great as that of the chassis frame. The latter was estimated by the method of (11) assuming no warping inhibition. Even if complete warping inhibition is assumed in the joints of the chassis-frame (say by the floor-panels acting as large gussets) the ratio of stiffness would be 500:1 and it can be safely assumed that, for metal bodies of the type analysed, all the torsion is absorbed by the body shell.

To confirm the assumption that the roof and floor act as shear panels, Fig. 9.19 shows the polystyrene model in torsion. The lower model shows the sides connected by continuous sheets at floor and roof level, front and rear bulkheads representing the actual pillar stiffnesses. The

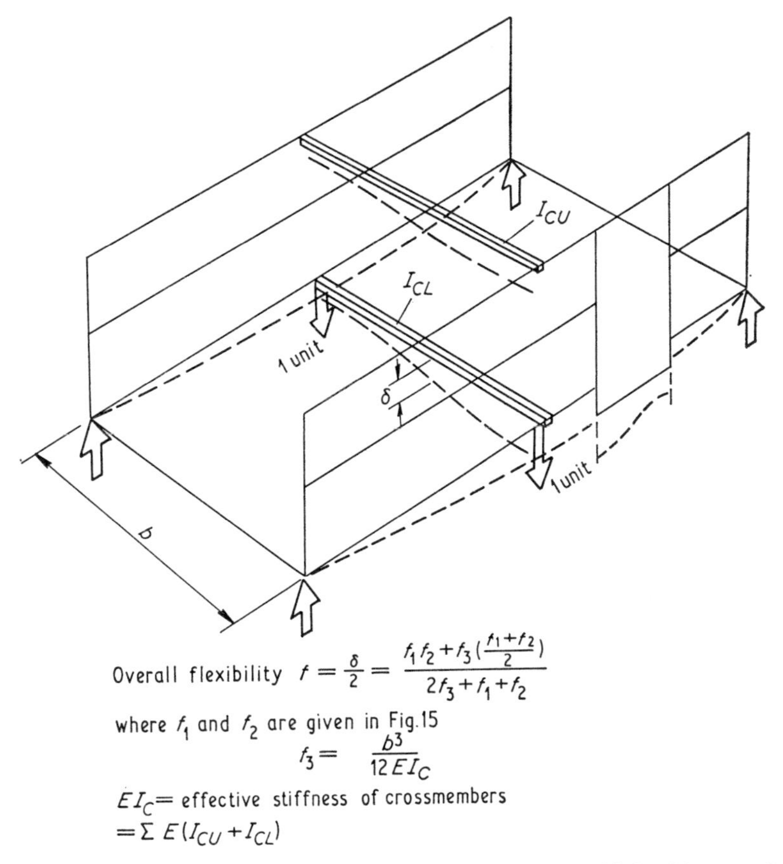

Fig. 9.17. The effect of the rings connecting the two vehicle sides on the overall flexibility

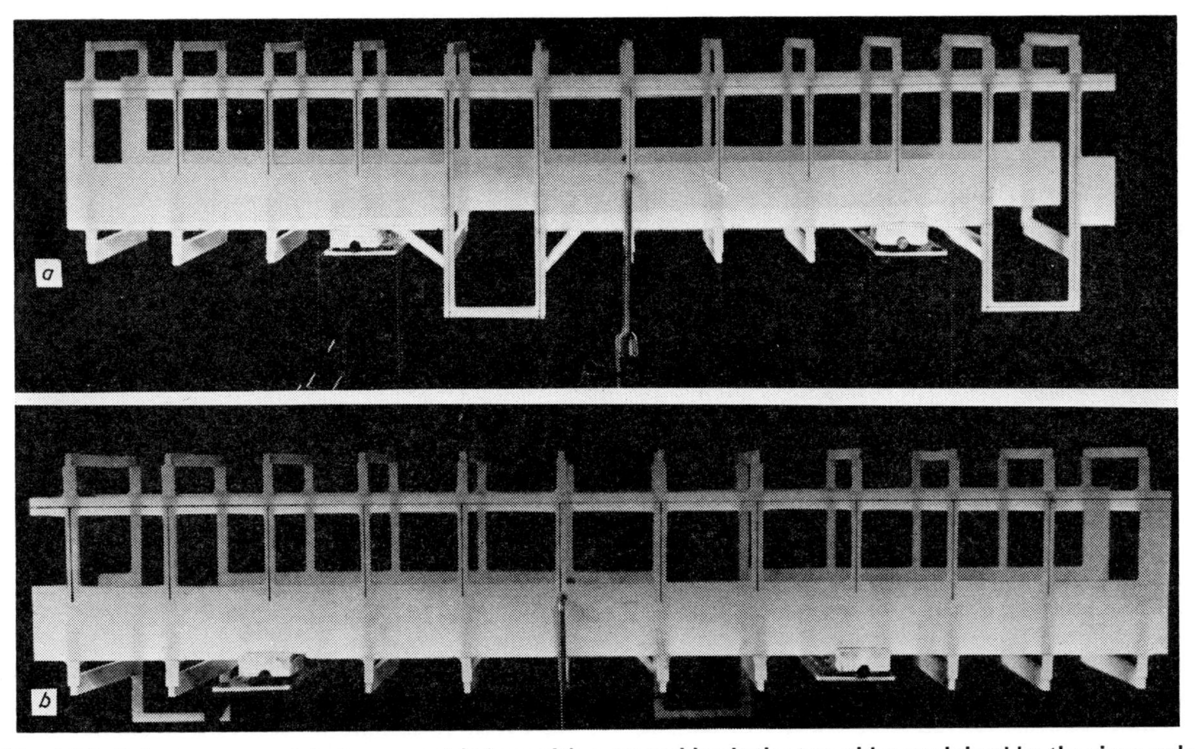

Fig. 9.18. Polystyrene model of asymmetric bus with a central load: the two sides are joined by the rings only

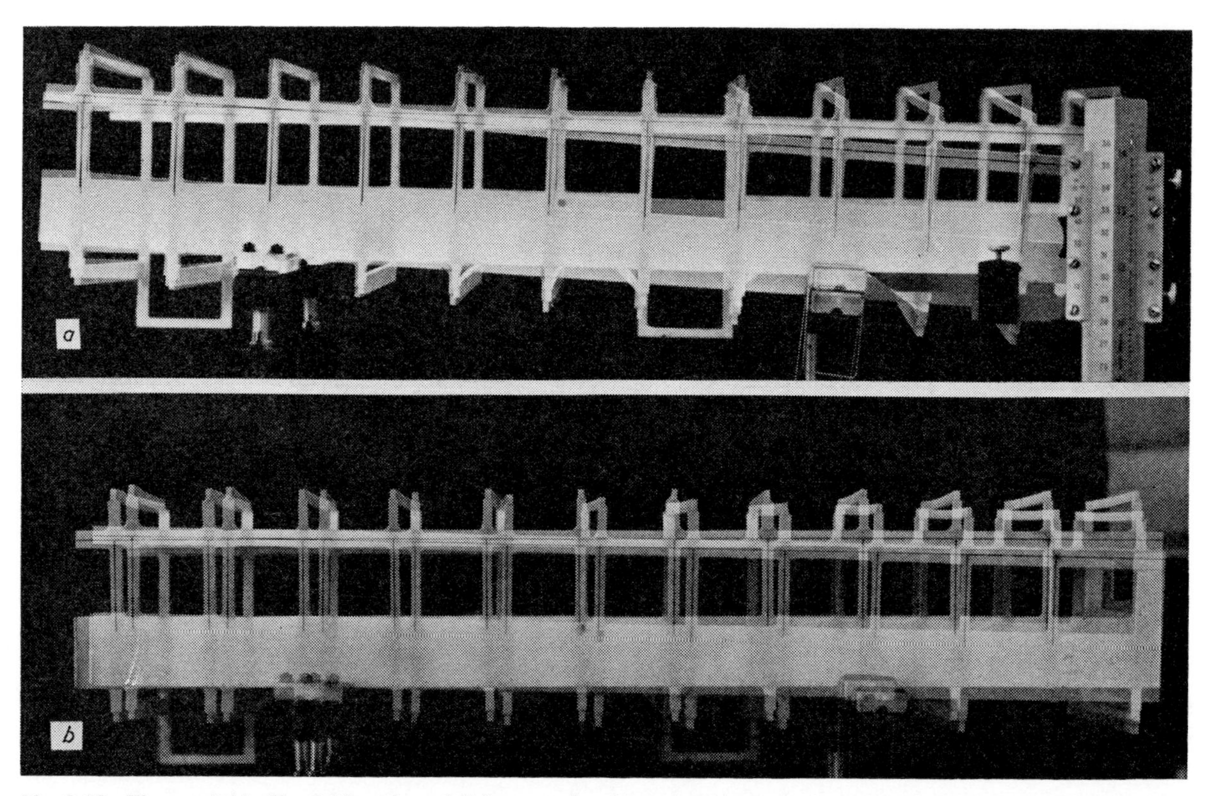

Fig. 9.19. The model in Fig. 9.18 under a 1 lbf corner load (top) and, at the bottom, the same model with sides joined by roof and floor shear panels and front and rear bulkheads; under a 5 lbf corner load

torsion load has been increased five times over the load used at the top where the only connection between the sides is provided by the rings. The vertical deflection of the load point (right-hand near side in the photograph) is seen to be reduced in spite of the increased load.

The torsion test rig consisted of vertical supports at three of the axle support points and a vertical load applied at the fourth point. Sufficient horizontal flexibility was provided at two of the supports to prevent constraint of the torsional deflection. There was some horizontal instability in the test rig which made measurement impossible but the vertical deflection of the support point (left-hand support in the photograph) can be seen to be negligible.

CONCLUSIONS

The work presented has demonstrated the use of simple theory to estimate the proportion of the total load carried by the body structure of a particular bus. Before the empirical constants used in these calculations can be used elsewhere with any real confidence, it will be necessary to analyse several other body–chassis combinations. Since this paper has only demonstrated consistency between simple theory and more sophisticated theory using a computer and simplified models, it is necessary to carry out stiffness measurements on full-scale vehicles before the methods presented can be used for detailed stress analysis.

Even within the framework of the present work there are some unsolved problems, e.g. the apparent 'shortening' of the window-pillars at the front of the vehicle in the computer-plotted side-frame deflection diagram in Fig. 9.16; and the effect of carrying the torsion-shear flow over the door-opening.

In spite of these limitations it is hoped that the methods outlined will enable design calculations to be carried out on chassis-mounted bus bodies with the same confidence as those for integral vehicles.

When the proportion of the overall load carried by the various parts of the structure is known, the formulae in Fig. 9.7 can be used to calculate the local stresses in the critical structural members.

ACKNOWLEDGEMENTS

The results in this paper supersede those of Hochberg (10) and Tidbury (12) (which carried some results from (10)), but the pioneering work of J. Hochberg in this field must be acknowledged. In his present capacity as chief engineer of the Ha'argaz Company he has further contributed by the provision of information on the structures of the vehicle analysed.

Much of the work reported here, including the manufacture and testing of the polystyrene models, was carried out by A. J. Campos as part of a Technical Essay for submission for a D.Au.E.

APPENDIX 9.1

Notation

b Width of bus structure (Fig. 9.17) (in Fig. 9.8 it is strictly the distance between the load inputs to the structure).

e Wheelbase (Figs 9.8 and 9.15).

f Overall body flexibility (bending or torsion).

f_1 Bending flexibility of the body-side without a door expressed as a central deflection per unit load when the side is supported at the axle centre lines (Fig. 9.15).

f_2 Bending flexibility of the body-side with a door opening defined as for f_1 (Fig. 9.15).

f_3 Shear flexibility of the structure joining the two body-sides. (Either complete rings or upper and lower fixed ended cross-members) (Fig. 9.17).

f_i Torsion flexibility of the whole body when only one frame (the ith frame) is allowed to deflect, the other frames being assumed infinitely rigid.

h Height between floor and roof planes (Fig. 9.7).

h_1 Height between waist-rail and cant-rail or roof plane (Figs 9.7, 9.8 and 9.10).

h'_1 Windscreen depth in its own plane (for a vertical windscreen $h_1 = h'_1$) (Fig. 9.8).

l Length of cant-rail (Fig. 9.8).

l_1 Length between front axle centre line and the forward door-pillar (Fig. 9.15).

l_2 Width of door-opening (Figs 9.7 and 9.15).

l_3 Length between rear axle centre line and the rearward door-pillar (Fig. 9.15).

q Shear flow between the structural planes of the body when in pure torsion (Fig. 9.8). (Fig. 9.2 shows an alternative method of defining q which is not generally accepted.)

t Notional thickness of the thin-walled tube carrying the torsion between the side-frames (Fig. 9.8).

A Cross-section area of bus viewed from the side (Fig. 9.8).

E Young's modulus.

I_{CL} Second moment of area of a floor cross member (Fig. 9.17).

I_{CU} Second moment of area of a roof-bow (Fig. 9.17).

I_L Second moment of area of door-sill (Figs 9.7 and 9.15).

I_U Second moment of area of cant-rail over door (Figs 9.7 and 9.15).

I_r Second moment of area of the rth window-pillar (Fig. 9.8).

I'_s Second moment of area of side structure below the cant-rail (Fig. 9.15).

I_s Effective second moment of area of complete side structure where there is no door-opening (Fig. 9.15).

M Bending moment at centre of door-opening (Fig. 9.7) or, bending moment at the base of a portal (Fig. 9.10).

M_D Torsion moment applied to the structure about a fore and aft axis.

M_L Bending moment at end of door-sill (Fig. 9.7).

M_U Bending moment at end of cant-rail over door (Fig. 9.7).

M_r Bending moment at the base of the rth window-pillar under various assumptions (Fig. 9.10).

M_{pr} Bending moment at the base of the window-pillar due to vehicle bending (Fig. 9.7).

M_{tr} Bending moment at the base of the rth window-pillar due to torsion (Fig. 9.8).

P Direct load in door-sill and cant-rail due to bending (Fig. 9.7).

Q Shear at door-opening centre line (Fig. 9.7) or, half side force carried by a portal (Fig. 9.10).

Q_L Shear carried by door-sill $\left(=\dfrac{I_L}{I_U+I_L}Q\right)$ (Fig 9.7).

Q_U Shear carried by cant-rail over door (Fig. 9.7).

Q_r Shear force transmitted by the rth window-pillar (Fig. 9.10).

Q_{pr} Shear force transmitted by the rth window-pillar due to bending (Fig. 9.7).

Q_{tr} Shear force transmitted by the rth window-pillar due to torsion (Fig. 9.8).

R External forces applied to rigid end bulkheads in the Brzoska analysis (Fig. 9.2).

$\bar{R}$ Forces applying torsion to the outside edges of the bulkheads in the Brzoska analysis (Fig. 9.2).

APPENDIX 9.2

Derivation of formulae

Loads in a side-frame due to bending (Fig. 9.7)

The bending moment at the centre of the door-opening (which may be anywhere along the side) is read off the bending moment diagram as M. The tension in the cant-rail and compression in the door-sill will be equal and opposite and given by $P = M/h$.

The tension force will be transmitted to the rigid section below the waist-rail by bending in the window-pillars and each pillar will transmit a force in proportion to its stiffness, i.e. if all pillars are of the same material:

$$Q_{pr} = \frac{I_r}{\sum I_r} P$$

For the structure illustrated in Fig. 9.7 the summation of I_r will be over six pillars for the section between the doors but only over two pillars for the rear of the bus. The constant of $\frac{2}{3}$ in the formula for the moment at the base of the pillar is discussed in the text and in Fig. 9.10.

In the case of the transfer of shear over the door the total shear force Q is again distributed between the cant-rail and door-sill in proportion to their bending stiffnesses. If each of these shear forces were acting on cantilevers as shown in Fig. 9.7*b*, the bending moments at the ends would be $M_U = 0{\cdot}5Q_U l_2$ and $M_L = 0{\cdot}5Q_L l_2$, but Erz introduces a factor of 1·5 in each case to allow approximately for the complete ring effect giving the formula quoted in Fig. 9.7 of $M_U = 0{\cdot}75Q_U l_2$ with a similar value for the bending moment at the sill of $M_L = 0{\cdot}75Q_L l_2$.

Bending flexibilities of side-frames (Fig. 9.15)

Side with no doors.

The deflection per unit load at the centre of a simply supported uniform beam is $l^3/48EI_s$, where I_s is the effective second moment of area of the complete side. An upper value for I_s could be obtained by considering the beam as having the full depth of the side including the cant-rail, waist-rail and other longitudinal members at floor level, giving a value of 31 905 cm^4. This assumes that the window openings transmit shear as efficiently as the continuous panelling below the waist-rail. If it is assumed that window-pillars are ineffective in transmitting shear a lower value for $I_s = I'_s$ would be obtained by including only the waist-rail as the upper flange and the floor level longitudinals (and any skirt runners) as the lower flange. $I'_s = 5700$ cm^4.

In view of the large difference between these values the computer result illustrated in Fig. 9.16 was used to find an arbitrary factor which can be applied to I'_s to give the same deflection using the simple formula

$$\delta = \frac{e^3}{48EI'_s} = 2\cdot88\times10^{-4}\ \text{cm/unit load}$$

$$\delta_1\ \text{(computed result)} = 1\cdot67\times10^{-4}\ \text{cm/unit load}$$

For

$$\delta_1 = \frac{l^3}{48EI_s},\quad \frac{e^3}{I'_s} = 1\cdot7\ \text{(approximately)}$$

This result will clearly only apply to vehicles of this type and the ratio (1·7) would be reduced for larger window-pillar spacing and increased for closer spacing. It is hoped that it will give a good first approximation as do the Erz constants.

Side-frame with doors.

This is treated as two sections of beam with the same effective second moment of area as the side without doors (I_s) with a section over the door-opening subject to the shear deflection only which is added directly to the bending deflection. As for other approximations in this analysis recourse is made to comparison with model and computer results to justify the addition of the whole of the shear deflection over the door to deflection at the centre of the side.

Application of the unit load method to the two sections of beam l_1 and l_3 joined by a rigid section over the door (l_2) gives the following integrals:

$$\delta_{\text{beam sections}} = \int_0^{e/2}\frac{x^2}{4EI_s}\,dx+\int_0^{l_3}\frac{x^2}{4EI_s}\,dx+\int_{l_2+l_3}^{e/2}\frac{x^2}{4EI_s}\,dx$$

$$= \frac{1}{12EI_s}\left[\frac{e^3}{4}-(l_2+l_3)^3+l_3{}^3\right]$$

The deflection due to shear over the door can be considered as the deflection of a fixed ended beam when one end is allowed to move relative to the other normal to the length of the beam, i.e.

$$\delta = \frac{Fl}{12EI}$$

where F is a normal force applied in opposite directions at each end.

In this case F = half the unit load since the shear force on each half of the side frame is half the centre load and $I = I_U+I_L$.

Therefore deflection in shear over the door-opening would be

$$\delta = \frac{l_2{}^3}{24E(I_U+I_L)}$$

It has already been noted that an empirical factor of 1·5 on the bending moment at the end of the beams makes an approximate allowance for the fact that they are part of a ring system. This factor becomes squared when applied to the deflection. Therefore

$$\delta_{\text{door-opening}} = (1\cdot5)^2\frac{l_2{}^3}{24E(I_U+I_L)}$$

and the flexibility f_2 of the whole side is $f_2 = \delta_{\text{beam sections}}+\delta_{\text{door-opening}}$ as in Fig. 9.15.

Overall bending flexibility of an unsymmetrical body

As shown in Fig. 9.17 two unit loads are applied to the centre of the vehicle, the overall flexibility will therefore be half the deflection given by these loads at the centre ($\delta/2$). Due to the shear stiffness of the structure joining the two sides the force acting on each side-frame will not be a unit force but F_1 and F_2 where $F_1+F_2 = 2$. Let δ_1 be the deflection of the side with no door due to a force F_1 acting on it and δ_2 be the deflection of the side with a door due to F_2. Then the shear deflection $\delta_3 = \delta_2-\delta_1$. If the shear flexibility is f_3, $\delta_3 = (F_1-F_2)f_3$.

But $\delta_1 = F_1f_1$ and $\delta_2 = F_2f_2$. Therefore

$$F_2f_2-F_1f_1 = (F_1-F_2)f_3$$

or

$$F_1 = \frac{2(f_3+f_2)}{2f_3+f_2+f_1}\quad \text{since}\quad F_2 = 2-F_1$$

and

$$F_2 = \frac{2(f_3+f_1)}{2f_3+f_2+f_1}$$

The deflection at the centre will be the mean of the deflections at the sides. Therefore

$$\delta = \frac{\delta_1+\delta_2}{2} = \frac{F_1f_1+F_2f_2}{2} = \frac{(f_3+f_2)f_1+(f_3+f_1)f_2}{2f_3+f_1+f_2}$$

Therefore

$$\text{Overall flexibility}\ f = \frac{\delta}{2} = \frac{f_1f_2+f_3\left(\dfrac{f_1+f_2}{2}\right)}{2f_3+f_1+f_2}$$

Torsional flexibility of a body structure

Referring to Fig. 9.8 the twist of the complete structure due to a unit value of M_D can be found by considering the deflection due to the non-rigid members in one side only.

The bending moment at any point along the rth window-pillars will be $\frac{2}{3}Q_{tr}x_r$, where Q_{tr} is the shear force acting on the rth pillar due to a unit value of M_D, i.e.

$$Q_{tr} = \frac{I_r}{\sum I_r}\frac{el}{2Ab}$$

The angular deflection due to this pillar will be, by the unit load method,

$$\theta_{\text{pillar}} = \frac{1}{EI_r}\int_0^{h_1}\left(\frac{2}{3}\right)^2 \frac{I_r}{\sum I_r}\left(\frac{el}{2Ab}\right)^2 x^2\,dx = \frac{e^2l^2I_rh_1^{\ 3}}{27A^2b^2E(\sum I_r)^2}$$

The angular deflection due to all the pillars will be the sum of the deflections due to the individual pillars, i.e.

$$\theta_i = \sum \frac{e^2l^2I_rh_1^{\ 3}}{27A^2b^2E(\sum I_r)^2}$$

Since all the terms except I_r inside the summation are the same for all pillars we can write

$$\theta_1 = f_1 = \frac{e^2l^2h_1^{\ 3}\sum I_r}{27A^2b^2E(\sum I_r)^2} = \frac{e^2l^2h_1^{\ 3}}{27A^2b^2E\sum I_r}$$

as in the text.

APPENDIX 9.3

REFERENCES

(1) BRZOSKA, Z. 'Basic problems in the statics of self-supporting vehicle bodies', *Archwim Budowy Maszin*, Vol. 2, No. 4, 1955, Vol. 3, No. 1, 1956 (in Polish).

(2) MICHELBERGER, P. 'Die Untersuchung von Autobussen mit Bodenrahmen oder Fahrgestell auf Verdrehung', *Actina Teknica* Vol. 35, 1961 (in German).

(3) MICHELBERGER, P. 'Wirkung der Turoffnungen auf das Kraftespiel der Omnibuskarosserien', *Periodica Polytechnica* (Budapest) Vol. 6, No. 2, 1962 (in German).

(4) MICHELBERGER, P. 'Das Kraftespiel einer Infolge der turoffnung unsymmetrischen Omnibuskarosserie mit elastichen Quertragern', *Periodica Polytechnica* (Budapest) Vol. 7, No. 3, 1963 (in German).

(5) ERZ, K. 'Uber die durch Unbenheten der Fahrbahn hervorgerufene Verdrenung von Strassenfahrzeugen', *A.T.Z.* No. 4; No. 6; No. 11; No. 12, 1957 (in German).

(6) PAWLOWSKI, J. *Vehicle Body Engineering*, G. H. Tidbury Ed. English Edition, Business Books, 1970.

(7) ALFREDSON, R. J. 'The structural analysis of a stressed skin bus body', *The Journal of the Institution of Engineers, Australia*, Vol. 39, No. 10–11, 1967.

(8) LEWIS, R. 'Structural analysis of an integral passenger coach by matrix force method', A.S.A.E. Thesis, 1966.

(9) WARDILL, G. A. 'Small computer procedures as tools for structural designers', *Body Engineering Symposium, Cranfield*, 1970.

(10) HOCHBERG, J. 'Stress analysis of bus bodies', A.S.A.E. Thesis, 1967.

(11) MARSHALL, P. H., ROACH, A. H. and TIDBURY, G. H. 'Torsional stiffness of commercial vehicle chassis frames', *XII Congress International des Techniques de l'automobile, F.I.S.I.T.A.*, 1968.

(12) TIDBURY, G. H. 'Integral structures for P.S.V.'s', *Automotive Design Engineering*, September, 1968.

Paper 10

AUTOMOTIVE DESIGN WITH SPECIAL CONSIDERATION FOR SAFETY IN INTERIOR DESIGN

J. E. Fallis*

INTRODUCTION

WHILE THE ENGINEER, for the most part, deals in facts, designers must be able to deal in, and correlate correctly, the aesthetic aspects. Designers are creative and imaginative people because our role in industry requires that particular talent above all. I am not suggesting we have the monopoly of creativity and imagination; of course not, it must, and does, exist in all fields of industry.

Like our engineering colleagues, we designers have very real practical limitations and constraints imposed on us. Clearly we could all design exotic and way-out cars, impractical if very original—all very fine for our egos and prestige—which would bankrupt the organization in months.

Webster's New Collegiate Dictionary offers as one of its definitions of design, 'the arrangements of elements that make up a work of art, a machine, or other man made objects'. I think that sums up the difference between what used to be the stylist and is now the designer. We no longer 'ice the cake', or hang on brightness and ornamentation, we start at the beginning and pursue a design right through to production.

It is clear that a designer, whether he is working in the automotive, air transport or industrial field, is not, as is often supposed, a foppish dilettante, tame artist or a hippy! The designer must have his feet planted firmly on economic ground with a business eye on the inevitable fashion movement and always looking for improvements. Amongst the improvements required not least is increased safety.

THE CHALLENGE OF SAFETY

To combine safety with engineering feasibility is a greater challenge and demands more creativity and ingenuity than we are given credit for. Most successful designers have an engineering background, at least to the extent that makes them appreciate engineering problems.

We also have to be aware of the myriad safety and legal problems which have to be anticipated and solved in the cars, vans and trucks which today are being designed for going on the road four or five years hence. You will probably be aware that there are many statutory safety requirements and, from my company's point of view, there are many more that are mandatory. Many of these are very sensible and of real value but of course there are some that need strengthening to make them useful.

Here we rely on experience and judgement—even on the crystal ball gazing of our Safety Engineering and Research Departments. They give us their forecasts of what we might expect and we add to these our own honest feelings of how we can improve safety. Now these are value judgements which could be right or wrong: but we try to interpret the findings on the assumption that our worst fears will be realized. These fears refer to the sheer economic challenge of what, more often than not, will be an additional feature of the product as our safety devices become more sophisticated.

Inevitably these are proving more and more expensive, but we believe that it is very difficult to put a price on safety and lives. Of course, the customer does not always see it that way—the customer always wants a bargain and is sometimes a little wary of paying more to keep himself alive or, at least, undamaged, in the event of an accident.

There is a great deal of publicity right now, making people a little more aware than they were before of the dangers of accidents and personal damage: we might even get them to wear the seat belts we fit for their safety. This publicity is sometimes rather bizarre; for instance, the air-bag is at this moment the big talking point.

This controversy has only recently hit the headlines but we have already anticipated it and are looking into this device, together with many others. Whether the air-bag

The MS. of this paper was accepted for publication on 15th March 1970.

* *Chief Designer, Interior Design, Ford Motor Co. Ltd, Engineering Research Centre, Laindon, Basildon.*

will solve everything remains to be seen. Personally, I don't think it's the cure-all that some people think it is.

Environmental safety

My colleagues and I—and let me stress this—would like to see the motorist take the initiative in preventing accidents, rather than rely on the designer to try to ameliorate the consequences. But we have always been safety conscious and the designs of motor vehicles have always reflected this. Rather than deal with safety in interiors, I would therefore prefer to concentrate on environmental safety; that is, creating an environment in which a driver is less likely to have an accident. Prevention is better than cure.

Some of the major aspects of this obviously concern the right environment for the driver inside his car. We have to place him in a position which allows him to concentrate fully on his job, which is piloting his car untroubled by extraneous influences.

Naturally, we do not want to turn the driver into a monk! It is still fun to drive cars. But he should have good visibility with full visual command of the traffic scene, the road ahead, to either side and to the rear of him. He must be comfortable—not so comfortable that he will fall asleep, but the comfort of an alert and active driving position without physical unease or distractions.

He must be able to handle and control the vehicle from that comfortable sitting position, efficiently and effectively. In simple terms: he needs good ergonomics; he must be able to reach the controls, use them properly, know what they do, be able to read and accept signals on his instruments quickly and efficiently, without the risk of mistake. He must also be happy in his environment which should provide a pleasing aesthetic mould. He must feel safe, good and know what he is doing. This all adds up to one thing: anthropometrics.

INTERIOR FITTINGS

It is blatantly obvious that prevention of an accident by these means is much better than additive cushioning after the event. Of course, we will continue to try to protect as best we can those who either are not sensitive to their environment or have been struck by someone else who is not.

Let us look at some of this additive cushioning. From an interior designer's point of view we have already made moves to incorporate the air-bag in a car; clearly it is at present very much a knife-and-fork job and you may have heard the results of some of the tests carried out in America—they were not very happy. If this system is adopted, I would hope that something less obstrusive and large will be developed which can be packaged more conveniently.

I personally dislike placing functional safety items in such a way that the driver or the passenger is constantly being reminded of the fact that they have to be protected. I think that is an environmental disease. It is not conducive to peace of mind to be constantly reminded of the possibility of accident, and, in this particular instance, the possibility of accidental triggering. This makes it doubly difficult for the designer to create the right, reassuring environment.

We are therefore planning to conceal the system as cunningly as we can without affecting its performance. This means further sacrifices, I am afraid, of glove compartment and other space. It will certainly add to the overall complexities of equipment in front of the driver and passenger.

I believe that, whilst the passive restraint system is not fool-proof yet, as the safety research engineers will tell you, it is still more reliable and effective than the air-bag, especially if we pursue the sophistication of the so-called safety-seat. This is fixed to the floor in such a way that it is totally achored and has within it integral safety belts and very often a means of 'whip-lash' prevention. You may recall that the Shelby *Mustang* had an inverted Y-type safety belt, coming down from the roll bar overhead, across the shoulders and fixed across the front of the occupant. I think it has many posibilities for real improvement.

Other safety features in the interior are well known: practical, sensible, injury-reducing, break-away rear-view mirrors, break-away window regulators, recessed handles, to mention just a few. These are normally used and there are hundreds more. We will continue our efforts to find better ways of making all these items in such a way that injury is reduced in the unhappy event of an accident.

A word perhaps on the collapsible steering wheel: this is clearly a major injury-reducing innovation which we have adopted on our production vehicles. It has suffered, like all innovations, from being made over-safe and we shall continue to learn how to improve it, both functionally and visually. I foresee some exciting developments shortly which, in addition to fulfilling the functional and safety requirements to legal standards, will also achieve neater and slimmer wheel and column, thereby improving visibility.

COMFORT AND ERGONOMICS

Now let us go back to environmental safety which I prefer because it prevents, rather than cures, accidents. We have talked of concentration, visibility, comfort, ergonomics, cybernetics and anthropometrics, but how do we achieve all these? We try to obtain independent judgements on whether our seats fulfil these needs. We have a panel of doctors and university men who have made special studies of the human anatomy with reference to car seating.

It is a complex subject. You can sit in a seat for two or three minutes and feel very comfortable; you might be quite unaware of the discomfort that could come, say, after 150 miles. There is also the problem of seats that look comfortable but are not. Naturally enough, we want to get the best combination of short-term and long-term comfort with looks. We want seats that are inviting to sit in but are still comfortable after 150 miles. Experts, of course, don't always agree: to achieve uniform recommendations is well nigh impossible, though, we listen respectfully and cull from their contributions a compromise that genuinely reflects the majority view. Apart from the sitting posture, we are also concerned with arresting sliding across the seats, reducing rolling and keeping the torso upright comfortably, relieving the strain on stomach muscles. All these factors create comfort, alertness and concentration.

Visibility is another requirement, and not only vision out of the vehicle; it includes the visibility of instruments and controls; and particularly the readability of instruments and identification of controls. You may know that there is no standard symbol for any knob or switch anywhere in the world!

We believe, and there are many others who agree with us, that there must be a common symbol system. We have submitted a series of symbols covering all the control knobs, and you would think it would be a simple task to standardize them, but it is proving very difficult. We are constantly trying to break down national barriers where someone wants to go it alone and there is always the 'not-invented-here' factor!

We have obviously got to get away from the use of words or letters—it has got to be symbols because we are now in a world-wide market. It can be done; we believe in what we are doing because it is so important for someone getting into a vehicle to know what his controls are telling him and which controls to use. Anyone who has driven a variety of vehicles will know the dilemma of trying to figure out how to work the heater, very often while he is driving, too. How do the switches operate, what do they do? This lack of standardization alone is a serious defect militating against safety. It is certainly as bad as any other and its remedy is one of the essential requirements for a safe environment.

THE FUTURE

No paper of this kind could be complete without a glimpse into the future. In my opinion instrument-panels could recede further and further away from the driver; after all, this is where most of the secondary injuries occur and the best solution is to remove the cause as far as is possible. The alternative is to pull the panel (or part of the panel) back towards the driver to reduce the distance of acceleration.

Whilst I don't want to isolate the driver from his passengers completely, I do think we are going to put him into a more prominent and exclusive position to permit him to do his task satisfactorily. He could well be seated in a higher position than the passengers; he will then get better all-round vision (look how the 'A' pillar has been whittled away in the past few years for instance).

In line with our driver-oriented approach, there will be some exciting developments in instrumentation. These are being made possible by the rapid development in, for example, electro-luminescence which can provide clear, concise and well-lit instrumentation. All of this yields better cybernetics—easier-to-read messages; less risk of mistake.

I can imagine instrument information given to the driver in three stages—perhaps all on one circular dial in front of him. The first stage of this electronic masterpiece will provide broad running data. Perhaps a progressive build-up of squares on a graph which will indicate roughly what speed is being achieved. Exact data are not really required out-of-town: I believe it is sufficient for a driver to know he is doing 45 to 50.

As soon as he comes into a built-up area, he will flick a knob and the dial alters to a conventional speedometer giving a more accurate message. For exact data, the third stage is brought in with another flick of the switch: the speed is then indicated digitally to the nearest 0·1 mile. This could be linked with an audio warning set to a given speed-limit. It is not inconceivable that a programmed voice could come in with a warning.

What we are anxious to do is to provide the driver with only the exact information he needs and that means the data must not be superfluous or distracting. Thus, too, the clock in the car would indicate only segments of time until the driver needed to know the exact time, perhaps to switch on a radio programme; he could also have his journey time computed automatically or a progressive calculation of average speed needed to make that vital appointment.

The combinations and permutations that could be obtained between speedometer, tachometer and clock with a flick of a switch are immense but the criterion for design is to feed the driver only the exact information, and only when he needs it.

Clearly, the control buttons and knobs will come away from the facia—they will be out of arm's reach anyway and would be potentially injurious or complicate the environment. They will have to be pulled back to within arm's reach and, for ease of use, embedded alongside the driver. He will learn to identify the control by touch and position and we hope that all manufacturers will, by then, identify the function of the controls with a common set of symbols, already referred to.

It is worth stressing this once again because we are anxious to gain world-wide acceptance of these symbols and we would welcome comment on them. In the long run standardization must be beneficial to everyone, but especially the driver.

Stereo radio, tape reproduction, air-conditioning or a sophisticated form of ventilation will be commonplace. I have no doubt that advances will be made in anti-dazzle, both from sun and other vehicles' lights; in suspension, ride, road-holding, everything helps towards the proper environment for driving. After all, despite what we sometimes think in the industry, the driver is more important than the car. Perhaps we get our priorities wrong sometimes!

For trims and fabrics we will develop better cloth, better vinyls and better materials generally. We have some in mind, but cannot announce them at this stage. We shall also, obviously, extend our use of soft materials and trim in side panels, armrests with additional padding, head-linings, all to help reduce the likelihood of hip, head or general body injury in an accident.

CONCLUSION

The message I am trying to convey here is that I believe, as regards safety, the environment so much outweighs anything else that it must have precedence. We have, of course, got to protect the unfortunates who do get involved in accidents and we will continue to do this, but our main effort is to pursue environmental safety. Cars get faster, perhaps more lethal because of it, and the only real answer is to prevent accidents.

Where they cannot be prevented the restraint systems will have to do the best they can. Safety is, and has always been, with us. In our endeavours we make every effort to make motor cars safer for people to drive to their work or to their enjoyment. I sincerely hope that others, in their responsible capacities as road makers, city planners and people like yourselves, will all put in a marked effort to reduce accidents. I hope everybody will take up this challenge of saving life and reducing personal injury from accidents on the roads.

Paper 11

CHASSIS FRAMES

D. W. Sherman*

Over the years the role of the frame in American passenger cars has changed greatly. More recently, the virtue of the frame with respect to collision damage and passenger safety is receiving increasing attention. After a short discussion of frame function, the author explores the evolution of the frame as influenced by major changes in other vehicle components. Special laboratory equipment for durability testing and analysis of structures is complemented by a machine for crushing vehicles and components to simulate the effects of collisions. This permits a step-by-step analysis and provides data for design improvement.

FRAME FUNCTION

Development over the years

The automobile frame supports the engine and body like a bridge; but because of its shallow proportions, compared to those of the body, the structural engineer is likely to view its use for this purpose as a little incongruous, and properly so.

The frame in the American car of the early 1920s did in fact perform rather like a bridge. Furthermore, in combination with the, at that time, solidly connected engine it formed a very competent torsion member for resisting deflection and taking the stress caused by uneven road-surfaces. In those days, however, car heights were such that the depth of the frame could be almost whatever was desired, and its rails were straight in the plan view. Car bodies were of composite construction—wood, steel and fabric—and not very good structurally. So the frame and engine supplied most of the car's structural strength.

Since then many changes have taken place and a number of what might be described as emergency situations have arisen from time to time. Information gained from analytical and test work for their correction has had a profound effect on the structural design of the American car, so that a brief review seems appropriate. First, however, the objective, with respect to the riding qualities of American cars, requires comment, not because it is something new but, rather, because certain of the situations discussed would not have arisen if the objective had been different.

Probably because of the long distances normally over relatively straight roads travelled by American cars, soft riding and quietness are important ingredients in the overall performance requirement. In the normally sized car, at least, hard riding, pitching, vibration and noisiness are unacceptable. It is recognized that tighter steering, stiffer springing and less rolling are advantageous in certain driving conditions, regardless of noise. But the American car performance has evolved so as best to suit American driving conditions.

With the advent, in about 1927, of rubber engine-mountings, isolating the engine from the frame, the vehicle's torsional resistance was sadly impaired and, besides, the engine itself became a massive vibrator from the standpoint of low-frequency disturbances initiated by rough road-surfaces. Thus there was a period when vehicle stability and handling were very much on the ragged side.

The first major step towards offsetting the detrimental effects of the resilient engine mountings was the adoption of 'X' type bracing in the frame and, for a time during the early 1930s, American cars were quite acceptable from the standpoint of structure firmness. However, the introduction of independent front-wheel suspension, while offering handling advantages, imposed new structural problems as did larger, softer tyres, higher speeds, and lowering of the body floors, with resulting reductions in frame sections.

During this period, still in the 1930s, the American car was much improved with respect to handling and high-speed performance but left much to be desired with regard to solidity and response to broken road surfaces. At that time, with losses in frame rigidity and increases in body rigidity, the body and frame each contributed about half of the vehicle's overall rigidity.

During the later 1930s considerable attention was given to the effect of resilience in the connections between the

The MS. of this paper was accepted for publication on 2nd December 1971.

* *Director, Engineering and Research Division, Parish Sales, P.O. Box 200, Taylor, Michigan 48180, U.S.A.*

frame and body. Non-metallic but hard shims had been used for some time to prevent squeak but distressing road noise became prevalent when the bodies became all-steel. To subdue this, the shims had been changed to rubber, with considerable gain in quietness. Increased softness and a sort of absorption of road roughness generally provided more pleasant ride with a sort of 'relaxed' feeling. However, the increased resilience at the body connections reduced vehicle rigidity in both bending and torsion, producing some unpleasant side effects. In general, the American car impressed one as being somewhat flexible and fragile.

Until that time all work on suspension flexibility had been concerned with the vertical plane, deflection being in the range of 100 lbf/in of wheel displacement. In 1938 most cars were plagued with lateral shake at the rear over rough roads, a most unpleasant condition. Rather crude measurements were made, therefore, of the lateral forces imposed upon the frame by the rear axle and, much to everyone's surprise, this was found to reach around 2000 lbf for straight driving at 30 mile/h on a city street of average roughness. Further investigation indicated that the geometry of the vertical displacement of one rear wheel in relation to the other was such that forces of this magnitude were quite logical and to be expected.

This led to the use of a cross-link tie between the rear axle and the frame, so as to utilize the body-mass to restrain repeated lateral displacement of the rear axle. Actually, the force input was thereby increased, but the lack of repeat action cured the lateral shake responsible for the study in the first place. Further work proved that body-attachment resilience in the lateral direction also had a significant effect on rear-end shake, and on the general feeling of the car as well.

Soft body attachment

Since those earlier days a great deal of work—mostly trial and error type—has gone into the subject of body-attachment resilience and this has been an important part of American car development. In 1957, a car was set up with extremely soft mountings to discover the relative movement between frame and body if largely unrestrained. The body sat upon the frame on 6 soft rubber pads, about 2 in thick, deflecting about $\frac{1}{4}$ in under the static body weight. Indicator cards and spring-loaded pens were used to trace movement between them.

The car was carefully driven on a smooth road, straight ahead at 30 mile/h, and allowed to strike one fairly bad pot hole on one side. It was then returned carefully to the laboratory at very low speed, where the cards were removed. The readings were found to be large at the front and small at the rear in the vertical plane, and conversely in the horizontal plane, as shown in Fig. 11.1.

In driving the car in this condition, it was found to be abnormally quiet; most pleasing. Also, small road disturbances were completely absorbed without any apparent reaction whatsoever, giving a most unusual, and again pleasing, impression. On rough roads, of course, the car was no good at all, shaking violently and giving every indication that it was likely to fall apart.

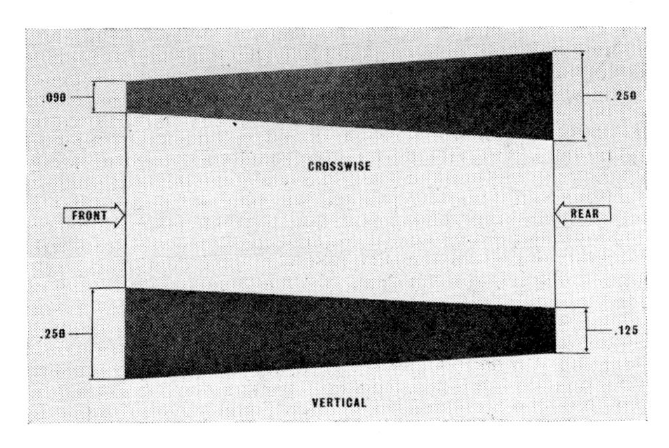

Fig. 11.1. Measured movement between frame and body, driving straight at 30 mile/h over one bump on one side only: rubber connectors were used

Rubber-bushed restraining links were then added at front and rear, so that each point could be completely restrained, either vertically, laterally or both, by tightening. A long series of experiments determined the effect of tightening. Interestingly, when the movement at one end was restrained by a certain amount in either plane, the movement at the other end increased by a like amount. In other words, the sum of the point motions remained the same. Or, we can say, average motion between frame and body remained unchanged.

Such changes improved the ride of the car at one end and worsened it at the other, but in all such combinations the car felt unusual, and extremely pleasing in certain respects. However, when both ends were tightened so that the total motion was reduced, the pleasantness began to disappear, the action becoming obnoxious. Finally, as tightening continued, the performance improved to that of conventional cars.

This series of experiments had rather important implications with respect to improvements in ride. Much work has since been done in attempts to secure equal softness and quietness while retaining good rough-road stability. This has not been fully accomplished, though the trend has been to softer inserts between frame and body.

Of considerable importance, furthermore, was the impression of much more severe force inputs as the connections between frame and body were tightened to reduce the average movement. Since the total movement was very small compared to that of the wheel suspensions, one would not have expected the force change to be very large.

To check this impression a weighted load-cell was dropped against the rubber mounting insert from increasing heights, until the deflection was equal to that measured on the vehicle when driven on reasonably rough road surfaces. Then the rubber was removed and the impact made metal to metal from the same height. The force was then found to be twice as great.

This would seem to be important information in considering the matter of separate frame and resilient body-frame connectors with respect to vehicle weight and durability.

As a final step, a very stiff and heavy special frame was installed in the test car to determine whether the rough road shake could thereby be controlled while retaining the soft, pleasant ride obtained with the very flexible body-frame connections. The stiffer frame had little effect, if any, and this led to the conclusion that the stiffness gain was offset by the additional weight. In other words, when the frame is freed from the body by resilient connectors, it must be regarded somewhat as unsprung weight with respect to the body. Increases in its mass, therefore, tend to prolong its action when excited, such prolongation leading to vibration build-up which is impressed on the consciousness of the passengers. In that case the ratio of frame-weight to stiffness is extremely important, functionally as well as economically.

THE MODERN FRAME

Since 1957, the continually reduced car heights have encroached upon the space available for the frame, so that in recent years the so-called 'perimeter frame' has come into common use as the only type for which space is available. Available space, in fact, has designed the frame, even though the concept was originated and thoroughly tested on the basis of forward planning long before space limitations made it a necessity (Fig. 11.2).

While this frame is quite efficient structurally, its severe configuration naturally reduces its structural support capability. This has had to be offset by improved body structure. In fact, for overall load-carrying, the modern American car body does not need a frame; a stub can be added to the front, as is done on some models, and the car will get along all right so far as transport is concerned. So why use a frame?

In the American car today the frame serves largely to isolate the body, insofar as performance is concerned. Attempts to save weight and cost by its elimination have largely failed and the loss of force reduction in the overall system, mentioned earlier, is believed to play a part in this, although this is not well understood and certainly hasn't been accurately evaluated.

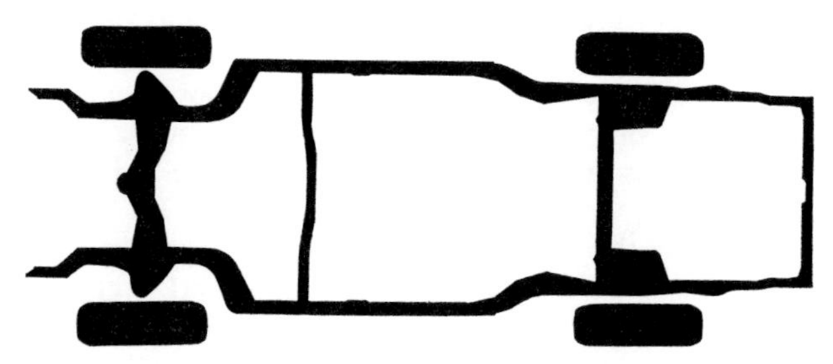

Fig. 11.2. A typical perimeter frame such as has been used in recent years to save space

Apart from this, the frame is believed to have a durability advantage for a given cost because it utilizes thicker sheet and more compact sections than is feasible for body structure. In other words, it's a simple, durable way to feed the wheel forces into the body-mass. Also, the separate frame seems to be convenient for carrying several body models.

The structural dynamics in cars is still something of a mystery; certainly it is complex, and probably much more so with resilient frame/body connectors than where the connection is solid. The ideal car, from the standpoint of ride, has yet to be achieved: it would have the softness and quietness of the American car, coupled with the anti-roll and steering of the English and European cars.

In the meantime, if one accepts that much remains to be, and will be, done, it seems quite likely that frames and frame/body connectors will be important factors.

Unlike some schools of thought I do not think it entirely proper to separate cars into large and small, with respect to this subject. But certainly, the small car would be better if softer and quieter without losses in other directions; and certainly the larger car would be better if more closely tied to the road without a loss of quietness and softness. In view of the known major effect of changes in structural rigidity and the tie between frame and body, it seems that this area perhaps offers the best opportunity for improvement, particularly since a mystery is also an opportunity.

COLLISION AND SAFETY

Regardless of current publicity, the design of automobiles and their components has always included contemplation of their reaction under unusual conditions such as collisions. The current emphasis, however, has made this subject almost into a trade of its own, as evidenced by one former Chief Engineer of an important works being delegated full-time to deal with it.

Collision energy absorption

When collisions reach a point of severity beyond that which can be cushioned by appurtenances such as bumpers and their attachments, the absorption of energy by destruction of basic structure comes into play. Here again the frame is important, it seems, if one gets down to basics. In this area we deal with energy conversion by means of metal-working and we must look for gains in efficiency by working more pounds of metal by greater amounts, within a given, limited, area.

In the design of structure, it seems therefore that thin and/or truss-like configurations, subject to failure by local buckling, should be avoided. Rather, we should aim for compact, fairly thick members that can be arranged so as to yield throughout. Because of the limitations of manufacture (spot welding), weight and cost (large, thin panels) body structures are mostly in the first category, basically light and thin, with local reinforcing members of similar nature. Frames, on the other hand, are more compact and thicker so that they would seem to be more capable of improvement, e.g. by the use of closed sections, bracing, gauge changes and suitable shapes.

For example, while frame configurations are somewhat limited by space, considerable latitude exists for modifications which enhance energy absorption and control rate and place of destruction; this can well add cost and probably will. But a very small local addition of metal can greatly increase the carrying capacity of a point that fails too early.

Special contouring, which is not so easily done on body

Fig. 11.3. A new crushing machine for studies of collision reactions of car frames

Fig. 11.4. A special machine for imposing road forces in the laboratory on a passenger car or light truck

Fig. 11.5. A similar machine to that in Fig. 11.4 for imposing road forces on heavy trucks

structure, can secure greater energy absorption. Collision tests to date show frame front-ends failing by buckling, with only a very small portion of the material being disturbed. Large gains can seemingly be made in this area; it's a case of determining just what is needed, and the least expensive way of doing it. If we could make every particle of metal yield in the first 12 in of the frame front-end, the effect on deceleration rate and vehicle damage would probably be startling.

Fig. 11.3 shows a crushing machine for studies of the collision reaction of structures. To the author's knowledge, its design is unique, offering advantages over other types currently in use. As can be seen, it consists of a central backbone, a power unit, and box section cross riggers for anchoring the vehicle. The ram is powered by hydraulic cylinders and guided by a central post. On the front are seven pressure-segments, each having a pressure-sensitive transducer for measuring force. The read-out equipment records the force from each, as well as the total.

The vehicle can be installed quickly for front, rear, or sideways crushing, and for angular crushing the entire power-unit, ram, cylinders, etc., are quickly lifted by crane and repositioned.

The crush-test is a method of observing and measuring, inch by inch, the behaviour of the members when force is applied to them. This makes possible the economic determination of means for increasing or decreasing their load-carrying capability, or for changing the character of their failure under load. In other words, it is a means of gathering design data, though not necessarily of improving design. In combination with dynamic testing, it is a means of improving energy absorption.

This machine was installed in March 1970 and is in constant use in a step-by-step, follow-your-nose process of finding out what can be done to absorb more impact with less destruction per foot, and to achieve a lower deceleration rate for the passengers.

LABORATORY DURABILITY TESTING

Fig. 11.4 is a special machine for reproducing and imposing road forces on passenger cars or light trucks in the laboratory. Fig. 11.5 is a similar but larger machine for trucks up to 80 000 lb G.V.W. The purpose of these machines it to check durability rapidly, while permitting observation, analysis and corrections to be made, and subsequent, quick re-testing.

They represent a combination of a special machine with hydraulic and electronic elements and with exhaustive and hard-won experience in how to use them for true simulation of road-service.

Special lightweight data-acquisition equipment for tape recording on-the-road movement of the wheel spindles (Fig. 11.6) is also used. The machine consists of tape-controlled hydraulic actuators for duplicating vertical wheel action. Along with this go linkages which duplicate side-loading as a function of vertical wheel movement.

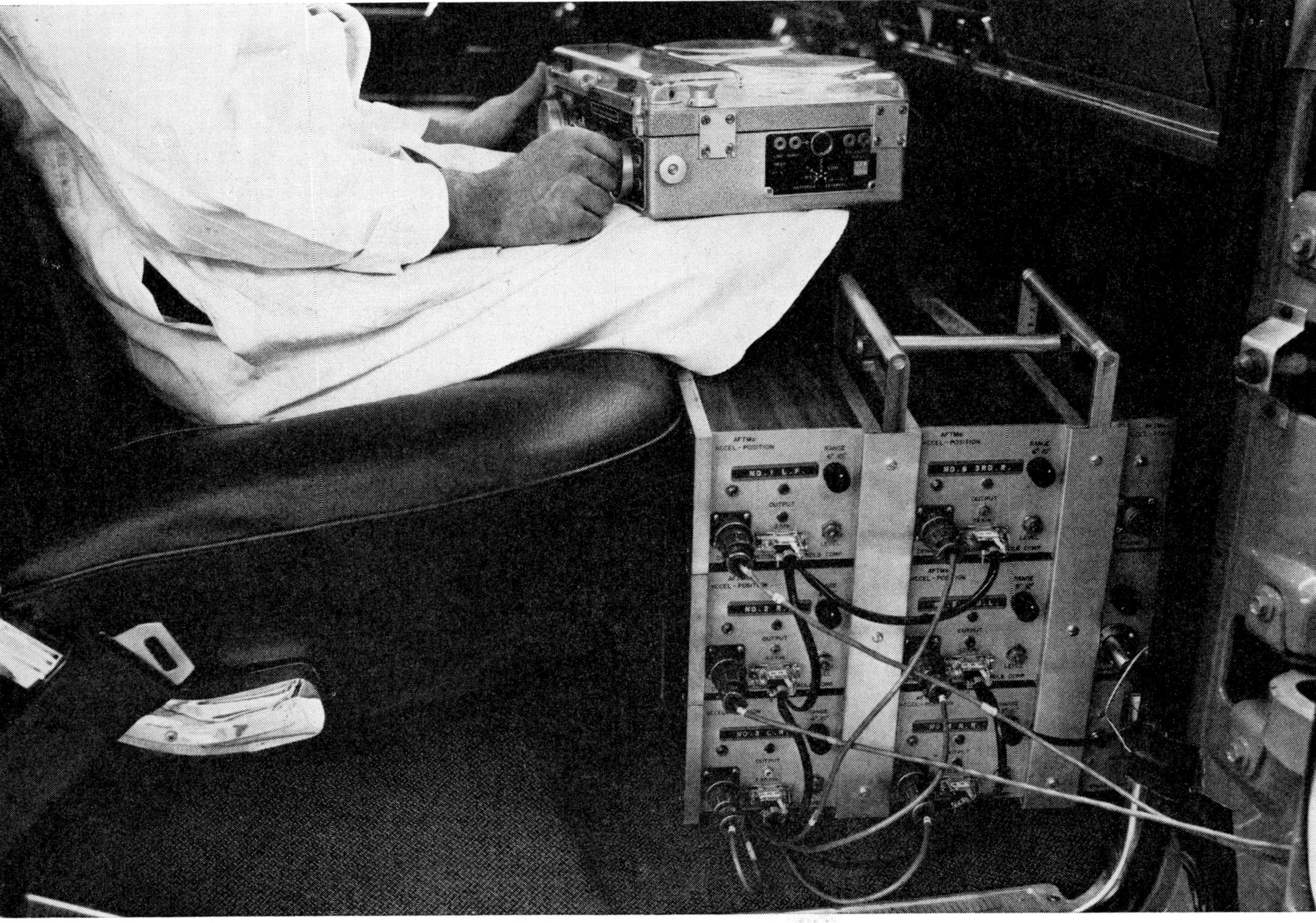

Fig. 11.6. Special lightweight equipment for tape recording the movement of wheel spindles on the road

Fig. 11.7. The control panel of the road-forces simulator shown in Figs 11.4 and 11.5

Hydraulic cylinders are programmed to introduce braking and acceleration forces at predetermined intervals. Fig. 11.7 shows the control panel.

This equipment reproduces field and service failures with unusual accuracy and in proper sequence. The tests are not accelerated by overload, but the net result is a speed-up because all extraneous delays are eliminated, as are such things as coffee breaks, and the equipment works 16 hours each working day. A total life-test takes 15 days to a month. Test work goes on continuously for the automobile industry.

In conclusion, it is hoped that by describing the evolution and function of the frame in the American car, and by acquainting you with the analytical work being carried on it, this paper will have contributed something useful for future development.

Paper 12

HUMAN FACTORS INFLUENCING CONTROL POSITIONS

G. R. W. Simmonds*

The ergonomic aspects of interior fittings is discussed with special reference to safety. International standardization of control positions and markings is an urgent need and more information is needed to help designers and stylists.

INTRODUCTION

A VERY LARGE number of factors govern the positions of the controls in a vehicle. A sizeable proportion of these influences involves people and as such could be termed 'human factors'. This paper does not attempt to cover all these. It does not deal, for example, with appearance or styling. Rather, it seeks to review those areas of study known as 'ergonomics' or 'human factors'. The main influences covered are reach, separation and layout of controls, and injury reduction.

In spite of this limitation, there are too many aspects and data to be covered in any detail by one paper. The aim is to review the sources of data and the methods of applying these data to vehicle designs and layouts. For a description of the general instrument panel design process, I refer you to the paper by Nissley and Elliot (**1**)†.

REACH

Variability

The need of a driver to reach his controls poses a problem largely because people differ in size. The use of percentiles to describe the distributions of these sizes is well known within the motor industry. However, there seems to be some misunderstanding of terms such as '95th percentile man'.

It is usual to record percentiles for individual dimensions such as standing height, sitting height or weight. Generally speaking, there is a correlation between these dimensions, but not a particularly strong one. Thus, a man who is taller than 95 per cent of his male colleagues is unlikely to have exactly the 95 percentile weight. This point is very graphically illustrated by Daniels and Churchill (**2**) who showed that, out of a group of 4063 men, none were even approximately 'average' on all the ten dimensions most relevant to clothing design. Their results are summarized in Fig. 12.1. A 95th percentile manikin is based on compromise dimensions.

Synthetic methods

When dealing with reach, then, it is far better to think in terms of '5th percentile reach' than the reach of a 5th percentile driver. Incidentally, it is of course the large man rather than the small woman who is likely to have difficulty in reaching car controls since his long legs encourage him to push his seat well back.

The principal methods in current use do not strictly measure percentiles of reach. These methods are synthetic in approach, that is, they attempt a mechanical simulation of drivers, based on the dimensions of major bones, etc. Even a barest appreciation of the anatomy of the human shoulder leads to the realization that such methods are likely to be crude (see, for example, Haslegrave (**3**)). However, surprisingly good results can be obtained with tools such as the one shown in Fig. 12.2.

The method which has been most widely used within the motor industry was produced by R. W. Roe (**4**). Measurements of functional reach show quite good agreement with this. Its major deficiency is that it substantially over-estimates reach from behind the steering wheel. Even so, its approach is interesting because it uses three 'levels' of reach for controls of differing importance. These are designated 'comfortable', 'extended' and 'leaning' reach.

The MS. of this paper was received at the Institution on 24th June 1970 and accepted for publication on 11th February 1971.
* *Principal Research Engineer, Ford Motor Co., Research Engineering Centre, Basildon, Essex.*
† *References are given in Appendix 12.1.*

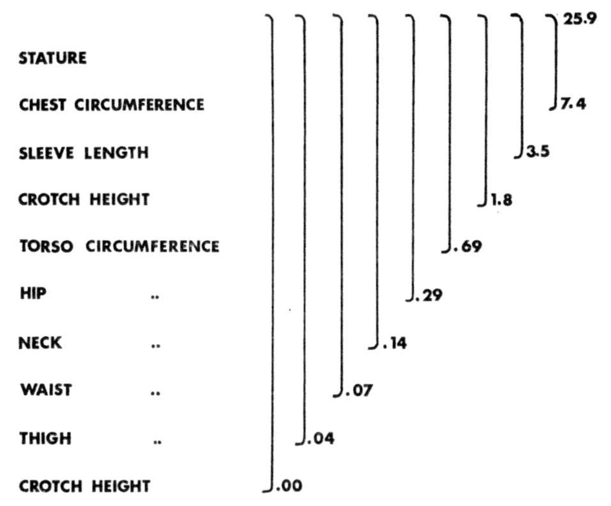

Fig. 12.1. Percentage of 4063 men approximately average in various groups of dimensions: none were average in all

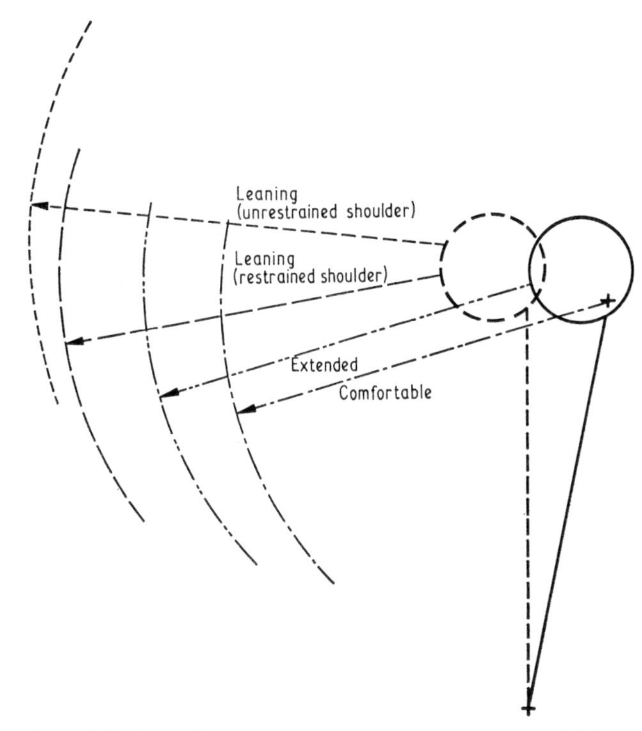

Fig. 12.2. Surprisingly good results have been obtained with this synthetic reach tool which crudely simulates human arm span

Such distinctions can provide drivers with the most acceptable compromise layout, but we do not have sufficient knowledge of the safety implications to warrant their inclusion in legislation. The principal authorities responsible for drafting safety legislation have indicated their intention of specifying minimum reach and it seems reasonable that these should relate only to the greatest reach possible, i.e. while leaning in a three-point belt.

For us in Europe, the most significant authority for safety legislation is the Economic Commission for Europe (E.C.E.) which seems likely to adopt a simplified synthetic approach. As with any simplification, a choice must be made between conservative or liberal errors. It is in the interest of both the drivers and the manufacturers that legislation should not adversely disturb the design compromise by unnecessarily eliminating acceptable designs, merely to ease the task of type-approval officers. The specification should be somewhat liberal.

Ergospheres

Traditional physical anthropological measurements of reach (e.g. Morgan *et al.* (**5**), Damon *et al.* (**6**)) are too general to be of much use to design engineers. The need is for package-oriented functional measurements. Such an approach has been taken by Chaffee (**7**). The basic approach stems from the aircraft industry (**8**), but Chaffee seems to be the first to have satisfactorily applied it to cars, and to have used female subjects. (Some elements of the approach as applied to cars are reported by Rebiffé and his collaborators (**9**).)

The best available anthropometric survey of adult civilians in the U.S.A., that by Stoudt *et al.* (**10**), was used to select a stratified sample. Each member of the sample pushed a series of knobs in selected positions as far as he could reach. Thus, a three-dimensional envelope of reach was found for each subject which was called the 'ergosphere' (Fig. 12.3).

Visual inspection yielded an epicentre within the ergospheres, termed the 'origin'. Intersections of individual ergospheres were distributed normally along lines through the ergosphere origin. Thus, 5th percentile positions could be calculated for representative lines through the origin. The surface through all these points is the 5th percentile ergosphere.

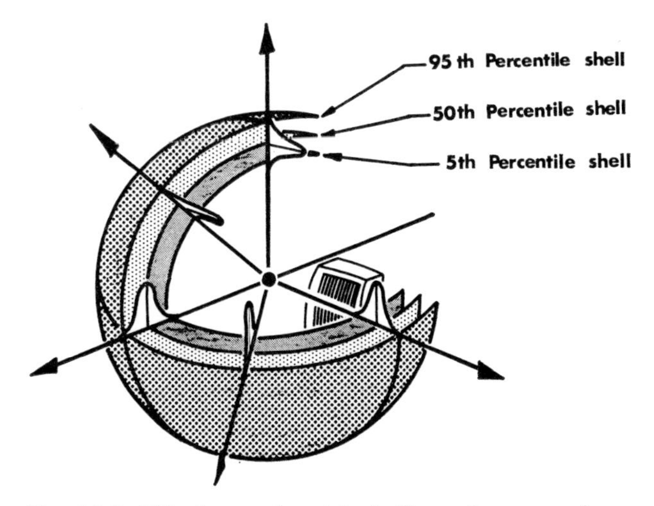

Fig. 12.3. This 'ergosphere' is built up from envelopes showing the functional reach for certain operation of the relevant percentiles

Further work is needed to establish ergospheres which are applicable to all package configurations. It is highly significant that the National Highway Safety Bureau (N.H.S.B.) of the U.S.A. has asked for proposals for research of this type. It seems likely that the most convenient form for practical ergospheres will be contour lines on transparent overlays, in the same manner as the eyellipses (11).

A feasible, and potentially useful, alternative to contour drawings involves the use of large electronic computers, possibly with graphic displays. Such an approach has already been applied to synthetic reach calculations by Boeing Aircraft Company, and in this country by Bonney and his colleagues of Nottingham University (12). This latter approach is still being developed.

Fitting trials

Fitting trials with a group of people in a seat mock-up provide a more cumbersome, but probably more exact, method of assessing the reachability of a particular layout. A properly selected, stratified sample of up to 40 drivers should be used.

Subjects are greatly influenced in their subjective assessment by the cars they normally drive, so care is needed in their selection. As with most reach surveys in vehicles, measurements should be made with the subjects wearing static seat belts which incorporate torso restraint. Care should be taken to standardize the degree of belt slackness. A convenient, socially acceptable and reasonably accurate method is for the subject to tighten the belt while it passes over the back of his fist which has, in turn, been placed on his sternum.

CONTROL SEPARATION

The variables

The main criterion to be met in establishing adequate control separation is the reduction of errors. The errors are false selection and inadvertent operation while using another control. The handbooks give detailed recommendations for control separation, and some American data have recently been re-published in (13) (14).

These data show the areas swept by large hands when operating various designs of knob. They are useful in that details are given for both bare and gloved hands as shown in Fig. 12.4. Additional separation is required for the operation of unilluminated controls at night.

In this respect shape or mode coding are of considerable help. By mode coding I mean the use of, say, rotary switches next to toggle switches. Shape coding has not been popular with stylists, perhaps they have been inhibited by the very unpleasing designs developed for aircraft. Nevertheless, different shapes of knobs are likely to have increasing potential as more and more controls are required (1).

The effect of the separation of the foot controls has been a favourite stamping ground for student experiments. Many have concentrated on the reduction in braking time, and it seems likely that reductions of the order of 0·2 s are possible by optimizing this aspect.

Systems criteria

It seems reasonable to digress at this point and look briefly at the relevance of such criteria as reaction time. Times are easy to measure, and reaction time studies have a long

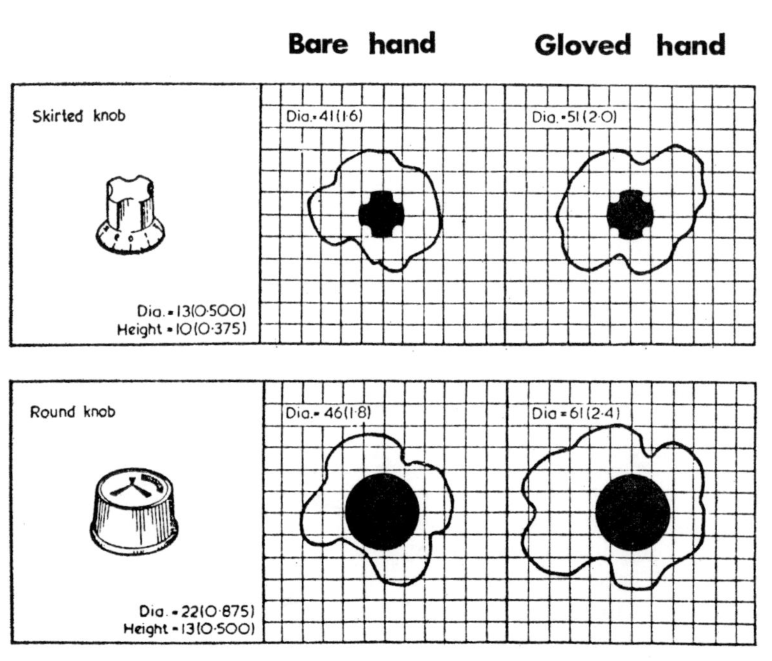

Fig. 12.4. The areas swept by the fingers when operating two kinds of control knob indicate the clearances required

tradition, but in the practical case of car design, they have little value in selecting an optimum solution.

Some workers have attempted to reduce these times to 'systems-relevant' criteria by translating them into distances covered at various speeds, some provide a conversion table, e.g. (15). I would suggest that it may be more appropriate to use the differences in the speed of impact if hard braking immediately followed the control action under consideration. It may perhaps be possible in the not too distant future to relate speed of impact to survival. It is possible to do so now in a crude way, for specific impacts.

A speed differential of almost 10 mile/h (4·4 m/s) is produced by emergency braking (0·9*g*) being delayed by 0·5 s. The extra half second may be well worth having for while survival of a 30 mile/h frontal impact with a rigid barrier is chancy, even if a lap and diagonal seat belt is worn, a similar 20 mile/h crash should be survivable (16).

LAYOUT

Optimum zones

When it has been established what is within the reach of 95 per cent of drivers, it is still not sufficient to scatter the controls at will within this volume. There are four factors which render certain zones more suitable than others, even though all are within reach. These factors are visibility, comfort, force and time-to-reach.

Visibility is more important for instrument displays but clearly it is also necessary for most controls, especially those not used frequently.

The first consideration is masking, especially by the steering wheel; this can be assessed for the driving population by using the extreme eyellipse tangents which show, superimposed, the differing zones of visibility for black-and-white and for the primary colours. Colours should not be used for identification purposes too near the periphery of vision. Generally, the more important hand controls should not be placed where they are masked from view, or towards the periphery of vision.

There are many data on forces that can be exerted by various groups of people in various directions and positions (5) (6) (9) (13) (17) (18) (19). These will certainly allow estimates to be made in many specific cases. However, experience indicates that they are not always adequate to answer specific queries on the suitability of particular controls, such as hand or foot brakes. Safe operation of these is very sensitive to the location of the point of application.

An example of the variation of pedal pressure at various angles is quoted on page 266 of Webb (19). Ford studies have shown that the maximum force that can be exerted on a handbrake may vary by as much as a factor of two between the positions of highest and lowest adjustment. Thus caution is needed.

Considerations of comfort are not normally of much importance for minor controls, unless excessive force or reach is involved. Comfort is important, however, in the location of the pedals and the steering wheel.

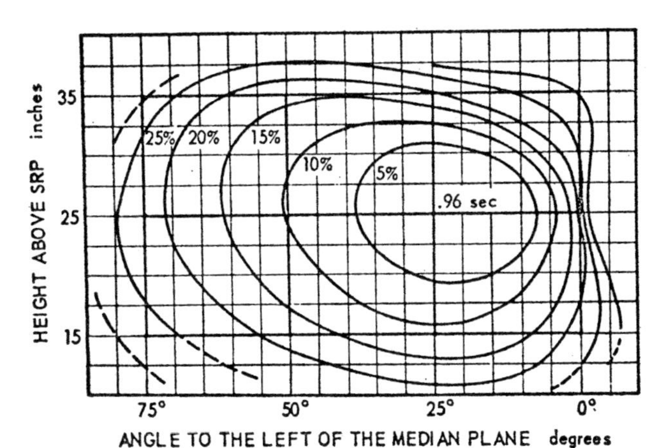

Fig. 12.5. The influence of position on the time needed to operate a toggle switch: the fastest time (0·96 s) is at an angle of 25° and 22 in above the H point

Comfort angles are used by many vehicle manufacturers in Europe, the U.S.A. and Japan. However, there seems to be a lack of unanimity. This is possibly due to different driving habits and different panel packages. Muscles are well able to adapt and what an individual finds comfortable is very likely to depend on the driving posture with which he is familiar. This may also explain the continuance of some of the incredible postures adopted by certain drivers, who seem to have a mammoth disregard for the position intended for them.

Some data are available on the times needed to operate various types of switch at various angles away from the centre line. The fastest (approximately 1 s) is likely to apply 560 mm (22 in) above the H point and 25° to the side (14) (19). At 75° from the longitudinal axis, the time will have increased by 20 per cent as shown in Fig. 12.5.

Layout criteria

It is clear that not all controls can be placed in the optimum position. What then are the criteria to be employed in selecting a good layout? There may not be a single optimum solution but there will almost certainly be a number of good solutions. There are four important criteria:

Frequency of use
Priority
Grouping
Standardization

It is self-evident that, if other things are equal, the more frequently used controls should be more conveniently placed. The problem lies in finding data which are representative over all driving conditions and for all drivers.

Even relative priority is not always easy to determine. It is clear that the steering wheel is a very-high-priority control and that the light switch should not be less conveniently placed than the cigar lighter. It is less clear whether the wiper should take precedence over the light switch.

In practice, however, it is probably not very important. On the other hand, it *is* important to avoid the confusion that might occur if, for example, the wiper and light switches were too close. Also, it is necessary to bear in mind that there is a tendency to reach short in emergencies so that the driver may operate a control nearer to him than the one intended.

Grouping may be in terms of function or of sequence of operation. In car design, it is probably better to group according to function. That is, controls relating to lights might be concentrated together in one area, and those for climate control in another.

The final criterion of standardization is one which instantly raises the hackles of stylists. Yet it is nothing new—we have standardized the locations of pedals for many years. Standardization is well accepted as a means of ensuring compatibility for mechanical parts such as light bulbs or trailer hitches. Man is flexible and adaptable, but not infinitely so.

With vehicle fleets, the extension of car hire and in two-car families there is an increasing need for a measure of standardization. This will not only be more convenient but will also increase safety for in an emergency there is a strong tendency to revert to an habitual mode of behaviour.

Standardization does not imply complete uniformity, it can be confined to the definition of zones in which certain types of controls are located. This seems to be the likely direction of future progress by the S.A.E. Recommended Practice J680 on layouts for trucks (**20**).

Other constraints

There are still other constraints. Displays such as speedometers and warning lights compete with the controls for prime panel space. Also, it is desirable that controls and related warning lights should be adjacent. Labels or symbols are required to identify controls in the increasingly complex layouts. These should be placed above or on the controls, so that they are not obscured by the approaching hand.

Conventions must also be observed regarding the direction of motion of controls. Thus, toggle switches should move down for 'on' in Europe. Other conventions are listed in the handbooks.

IMPACT SAFETY

There is a great lack of specific, reliable and relevant data on human tolerance to impact. The designer has little to go on other than U.S.A. Federal Motor Vehicle Safety Standard 201, which limits the acceleration/time characteristics of possible impacts. This represents an important constraint on design.

The safety requirements are antagonistic to some extent: accident avoidance prescribes controls within reach, injury reduction demands protrusions that are remote from the driver's head or knees.

CONCLUSIONS

There is a substantial need for more information which is relevant to car and truck design. More effort is required to produce further design aids for engineers and stylists. Not all aspects, however, can be covered with eyellipses, manikins and the like. More subtle tools are required and these will be created by stylists and engineers themselves if they receive the right training. The collection of the data will be the task of qualified ergonomists.

This does not mean, of course, that no useful data are available, far from it. In fact, a greater awareness of ergonomics by those with design responsibility would even now bring rewards to drivers.

What are the benefits to be gained? The prime benefit is a closer specification of the performance requirements of vehicles. This will not only lead to better vehicles but also will improve cost/effectiveness. A more exact specification of performance requirements presents the opportunity to put the money where it is most needed. The Japanese have organized themselves to do this (**18**) and we in Europe would also benefit.

ACKNOWLEDGEMENTS

The author wishes to thank the Ford Motor Company Limited for permission to publish this paper, and to recognize his indebtedness to his colleagues.

APPENDIX 12.1

REFERENCES

(**1**) NISSLEY, H. and ELLIOTT, J. 'Instrument panel design, the "control center" of the car'; New York, NY, (1970) SAE paper 700043.

(**2**) DANIELS, G. S. and CHURCHILL, E (1952) 'The Average Man?'; Wright Patterson AFB, Ohio, WADC TN 53–7.

(**3**) HASLEGRAVE, C. M. (1970) Study of Reach to Car Controls while restrained by a Lap and Diagonal Belt—Nuneaton, UK; MIRA Bulletin 11–15.

(**4**) ROE, R. W. (1966) Reach Distances with Upper Torso Restraint.

(**5**) MORGAN, C. T., COOK, J. S., CHAPANIS, A. and LUND, M. W. (1963) Human Engineering Guide to Equipment Design; New York NY, McGraw-Hill.

(**6**) DAMON, A., STOUDT, H. W. and MCFARLAND, R. A. (1966) The Human Body in Equipment Design; Cambridge, Massachusetts; Harvard University Press.

(**7**) CHAFFEE, J. W. (1969) Methods for determining Driver Reach Capability; New York, NY, SAE Paper 690105.

(**8**) DEMPSTER, W. T. (1955) Space Requirements of the Seated Operator, USAF, WADC Technical Report 55–159, 1955.

(**9**) REBIFFÉ, R., ZAYANA, O. and TARRIERE, C. (1969) Détermination des zones optimales pour l'emplacement des commandes manuelles dans l'espace de travail; Ergonomics 12, 913–924.

(**10**) STOUDT, H. W., DAMON, A. and MCFARLAND, R. Weight, Height and selected Body Dimensions of Adults; Washington DC, Public Health Service Publication No. 1000, Series 11, No. 8.

(**11**) ANON. Motor Vehicle Driver's Eye Range; New York, NY, SAE Recommended Practice J941*b*.

(**12**) BONNEY, M. C., EVERSHED, D. G. and ROBERTS, E. A. (1969) SAMMIE—A Computer Model of Man and his Environment; Paper presented to the Annual Meeting of the Ergonomics Research Society.

(13) ANON. Controls; Applied Ergonomics 1, 95–106.

(14) SHARP, E. D. (1969) Human Factors Considerations to the Design, Placement and Function of Vehicle Controls; New York, NY, SAE Paper 690459.

(15) BELZER, E. G. and HUFFMAN, W. J. (1966) The Quickness of Selected Right Foot and Left Foot Braking Techniques; Highway Safety Research Review. September 1966, 72–77.

(16) BOHLIN, N. I. (1968) A Statistical Analysis of 28 000 Accident Cases with emphasis on Occupant Restraint Value; Gotëborg, Volvo Technical Report.

(17) MURRELL, H. (1963) Controls and Instruments—Design Procedure; Automotive Design Engineering 2, October 1963, 70–74.

(18) OSHIMA, M. (1969) Ergonomics in Japan; Ergonomics 12, 701–712.

(19) WEBB, P. (ed.) (1964) Bioastronautics Data Book; Washington DC, NASA SP–3006.

(20) ANON. Location and Operation of Instruments and Controls in Motor Truck Cabs; New York, NY, SAE J680.

(22) ANON. Human Tolerance to Impact Conditions as related to Motor Vehicle Design; New York, NY, SAE J885.

(23) HUELKE, D. (ed.) (1970) Human Anatomy, Impact Injuries and Human Tolerances; New York, NY, SAE 700195.

(24) ANON. Layout of Panels and Machines; Applied Ergonomics 1, 107–112.

Paper 13

AUTOMOBILE BODY TESTING TECHNIQUES

W. R. Greenaway*

This paper is concerned with test techniques which can loosely be called static. These range from tests on complete body shells through those on sub-assemblies to tests on quite small details.

INTRODUCTION

Commencing with the complete bodies, regular tests, at the moment, are primarily concerned with stiffness. In general, strength is only important locally: if the structure of a mild-steel body is stiff enough it is strong enough. Both bending and torsional stiffness must be considered.

Since the provision of relatively rigid and accurately located mounting points for the running gear is of paramount importance, the body shell is attached to the test rig fixtures in the region of the wheel positions. Not only do these areas offer sufficient strength for loads to be applied to the rest of the structure but they are also the approximate points at which the service loads are applied to the body.

THE RIG

The Rootes rig shown in Fig. 13.1 is based on that used at Pressed Steel Fisher in order to obtain comparable results and thus be able to make use of previous information, obtained by Pressed Steel Fisher, on Rootes *Hunter* bodies.

The rig consists basically of two long, rolled-steel channels, bolted to the floor and spaced three feet apart. At one end of these are mounted two pillars (one of which is shown in Fig. 13.2) and a central hinge point with its axis parallel to the channels. To this hinge is attached a stiff beam, some six feet long, on which attachment devices suitable for whatever body is being tested are mounted. The forward end of the body is attached to this beam.

Other attachments on the beam can be used either to fix the beam rigidly in position for bending tests, or to take a turnbuckle, attached to a spring-balance mounted on one or other of the pillars, for torsion tests. The attachments for the rear of the body are mounted on a fixed beam fastened to the channels, as shown in Fig. 13.3.

The MS. of this paper was received at the Institution on 13th May 1971 and accepted for publication on 21st May 1971.

* *Senior Engineer, Product Engineering, Chrysler U.K., Coventry.*

For tests on both the *Hunter* and the *Avenger*, the front attachments presented no problems. The mounting for the McPherson strut makes an ideal attachment point. The fixtures used are angled to make the load pass through the centre of the road–wheel contact patch under static load.

On the *Hunter*, the rear attachment was also relatively simple—a dummy spring made of mild steel was mounted like the normal rear spring, with a pin at the front and a swinging link at the rear. This was attached to the fixed beam at the line of the rear axle.

The *Avenger*, however, presented a little more difficulty in that its rear suspension employs coil springs and trailing links, with the spring between the link and the rear underframe, the axle lying to the rear of the spring. Ideally, to maintain some comparability with the *Hunter*, a trailing link, with a solid strut replacing the spring and fixed at the axle line, would be the best solution. Unfortunately, whilst this arrangement is capable of transmitting compressive loads, it cannot transmit the tensile loads which arise during torsion tests, since there is no readily available means of attaching the strut to the underframe.

The rear attachments were therefore made at the bumper rubber mounting and angled to pass through the line of the axle. This means that, locally, the load paths into the body are unrepresentative of service conditions. However, since this is unlikely to have any significant effect on the overall stiffnesses of the body shell, this is relatively unimportant. The rear attachments are free to pivot about a lateral horizontal axis, thus avoiding any unnatural restraints due to too great rigidity. Fig. 13.4 shows an *Avenger* body under bending, viewed from the rear.

The deflections of up to 200 points under the various loading conditions can be monitored by standard dial-gauges, generally of one inch travel. At points where these deflections are expected to be large, two-inch travel indicators are used. These gauges are mounted on a separate scaffolding, and bear on small flat metal plates,

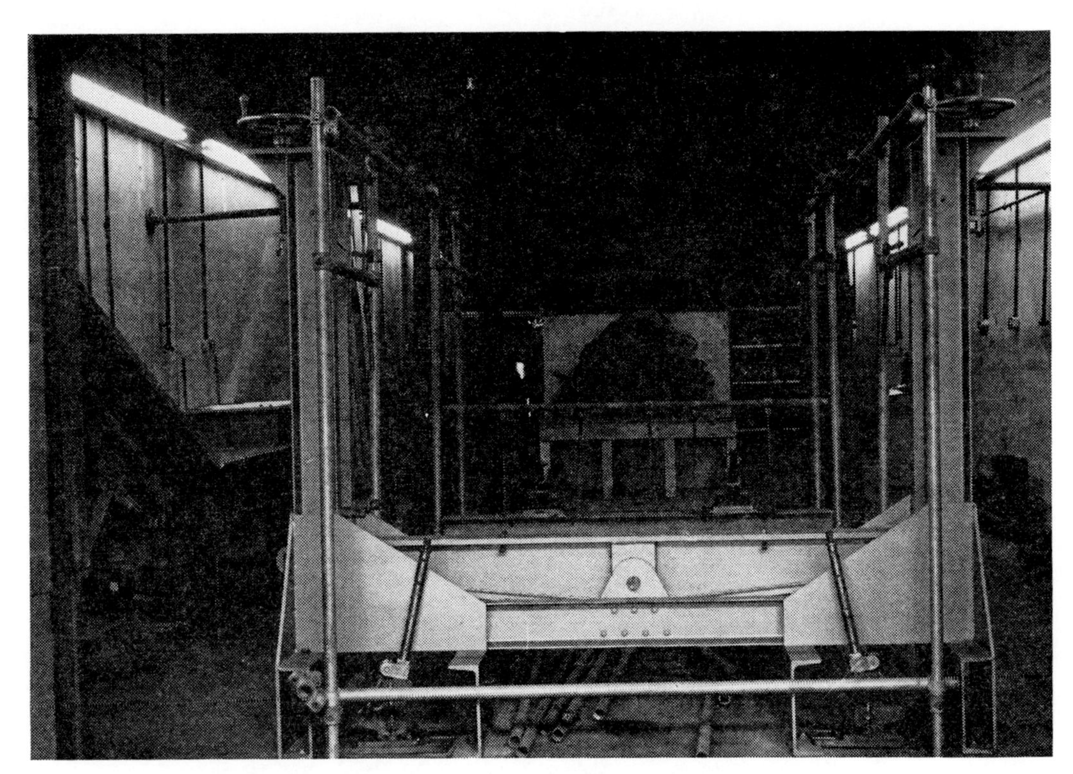

Fig. 13.1. General view of the stiffness rig from the front

Fig. 13.2. Pivoted beam and one loading pillar of the rig

Fig. 13.3. Attachments for the rear of the body are mounted on a fixed beam, fastened to the base channels

Fig. 13.4. The rear of the *Avenger*—seen here under bending—was attached at the bumper rubber mounting in line with the axle

Fig. 13.5. Simca body shell undergoing bending: a loading tree splits up external loads over seat and boot areas

attached to the relevant points on the body shell with Plasticine.

TESTING

Measurement and loading

At the beginning of each test all indicators are set at mid-travel. The initial set-up is rather a lengthy business but, since there are seldom any large distortions, the setting-up for each subsequent test is merely a matter of adjusting the zero setting of each indicator. The mid-travel position provides for deflection in either direction, depending on the nature of the test.

The usual procedure adopted is to carry out the torsion tests first. Prior to the first one in each direction a 'scrag' or pre-load is applied. This serves a dual purpose in 'settling' the structure and indicating the magnitude of the deflections which will be obtained during the test.

A torque of 3000 lbf ft is applied in one direction by increments and the readings of all the indicators noted. This procedure is repeated to check the reliability of the results. The whole procedure is then repeated with torque of the opposite sense.

The bending tests follow. Generally several conditions are simulated, including up to 3*g* turns with occupants and no luggage; the same with 600 lb of luggage; and any other combinations of design condition that are considered significant.

Usually sandbags provide the required loads. These are laid on the floor pan over the seat areas and in the boot, as required, care being taken to ensure an even distribution. Deflection indicator readings are taken at increments during each loading condition and each condition is repeated as a check.

An attempt was made on one vehicle body to correlate a method evolved by Simca with this sandbag method used by Pressed Steel. Simca use a 'loading tree', made up of suitable structural members, which splits up an externally applied load over the seat and boot areas (Fig. 13.5).

In many ways this seems preferable. However, considerably more effort is required in the preparation of the 'loading tree' which is unique for each type of body, and in setting-up. The bending tests themselves are much easier to conduct and do not involve the placing of heavy sandbags in rather awkward positions. This is a physically demanding activity and, unless great care is taken, can lead to disturbance of the body and, perhaps more important, of the structure on which the dial gauges are mounted (Fig. 13.6).

The results obtained from both torsion and bend tests are converted to deflections and fed into a computer which corrects for movement of the four mounting points and plots a series of deflection curves for each test.

Apertures

In addition to point deflection measurements, aperture distortion is also monitored during all torsion and bend tests. This usually takes the form of measuring the diagonals and centre distances of all apertures before and during tests. Centre-punch marks are made in the required positions and the distances taken off by trammels equipped with vernier adjustment and pointed probes. These are read off against a steel rule.

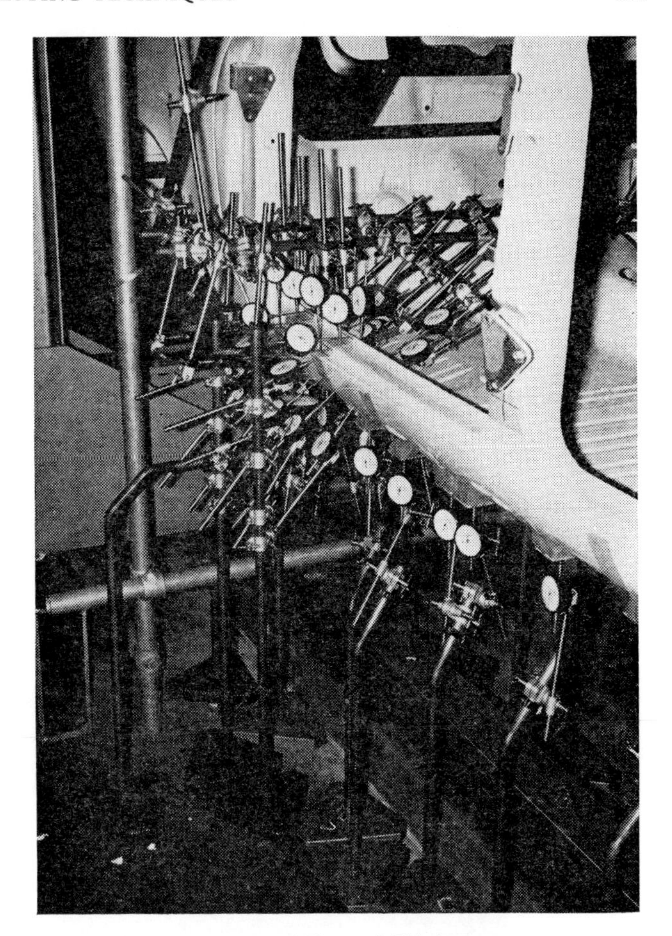

Fig. 13.6. Dial gauges at 'A' post/sill joint: great care must be taken not to disturb them

Whilst, in general, each measurement can be made to 0·010 in, the overall accuracy of any particular measurement of aperture distortion is of the order of $\pm$0·020 in. That this is not particularly close is of no great consequence when distortions remain small, since anything less than about 0·150 in is unlikely to be significant. However, although this method yields reasonably satisfactory results, it leaves scope for improvement. Consideration is being given to ways of doing this.

Attempts to use a dial gauge or vernier tape rigidly attached to one end of the dimension concerned are usually frustrated because this obstructs the loading of sandbags into the body, and because other interested parties are reluctant to permit damage to prototype body shells destined for use as road vehicles. This latter point also applies—perhaps more forcefully—to the mounting points.

The final tests performed on complete body shells are 'door drop' tests. These take the form of hanging weights

Fig. 13.7. Static rig for seat belt anchorage tests: distortion is of no consequence but dynamic strength tests would be of interest

Fig. 13.8

on the edge remote from the hinges when the door is open some 10°, and measuring the deflection. Particular note is paid to any permanent set which can arise from a number of sources and, if excessive, may call for more detailed investigation to pinpoint the source.

Detail testing

Before any design can reach the stage of body-shell tests, a number of detail tests may be necessary. These could be needed on almost any joint in the structure where some novel feature is to be introduced or where some departure from previous practice is contemplated; or on many other features, such as attachment points for running gear, engine mountings, and the like. These can be either strength or stiffness tests. The greatest difficulty is usually to decide how representative any test specimen has to be, not only of the proposed feature but also of the restraints applied to it. This is generally a matter of opinion.

In a typical example, say, a 'B/C' post attached to the sill joint, the post merely needs to be long enough to dissipate any local effects of end-fixing away from the joint. The sill, however, is to some degree affected by the floor and some of this has therefore to be included. This effect will depend largely on the type of loading to which the joint is subjected.

The other problem with this joint is whether the end-fixings for the two pieces of sill should, besides providing bending and torsional rigidity, also provide some degree of restraint against direct end-loads, or be free to float. Because of the practical difficulties in providing a fully floating attachment of this nature, Rootes have so far used completely rigid end-fixings for the sill.

The additional restraint built into the system is difficult to evaluate without an extensive investigation. The type of loading to which this joint could be subjected varies from bending coupled with shear applied to the 'B/C' post, either laterally or in line with the sill, to pure torsion in either sense.

The linear and torsional deflections of the joint would be measured by a number of suitably placed dial gauges. These could number up to 20, including the controls, but ten would give reasonable results. Other features have to be treated on their merits. Loading system and deflection measurement will depend on whether strength or stiffness is being investigated.

Safety requirements

Perhaps the most significant developments in body testing during the last few years have arisen from the need to comply with various mandatory safety requirements. At present, these are confined to anchorages for seats and seat belts, door hinges and latches, and impact behaviour of instrument panels. The legislation varies from country to country but is broadly similar.

Except for the instrument panels static tests have so far been acceptable; these merely require that the feature in question shall withstand some load without complete failure. Fig. 13.7 shows a rig for seat belts. Distortion or yielding are not of any great consequence and hence deflection measurements are rarely made. If at all, they are made for interest and not to meet mandatory requirements. This makes for easier setting up, as it is only necessary to ensure the correct load application.

Hinges and latches are tested in special fixtures adapted to suit a tensile test machine and there is little scope for variation in technique. Fixtures are fairly tightly defined by their function and the only variations tend to be due to the shape and nature of the test items. Again, it is only necessary to demonstrate that the latter are strong enough and any measurement of deflection is for interest only.

There are, however, moves to introduce dynamic testing of safety features. These will in all probability be on seat belt anchorages, doors, and roofs. In this instance there will obviously be considerable scope for variations in technique. So far the only clear requirement refers to door strength and this differs considerably from the original proposals.

Rootes approach to this test was to employ a hydraulic ram to drive the specific shape required into the door at the speeds originally suggested. Under the final requirements this is probably not the best approach but it can be made to perform the specified tests.

As for other dynamic tests, it remains to be seen what will actually be required but, on present knowledge, hydraulic loading systems seem to offer the best method.

Paper 14

THE EXPERIMENTAL INVESTIGATION OF BODY STRUCTURAL VIBRATION

M. Rodger*

To facilitate the analysis of the noise and vibration properties of prototype bodies the concept of 'mobility' is introduced, i.e. velocity-response/force input. Satisfactory results can be obtained with a single-excitation testing system, but this has certain limitations which can be overcome by a servo-controlled multi-point system now under investigation.

INTRODUCTION

SUCCESSIVE refinements to a system usually bring new problems and, just as the refinements become more and more sophisticated, so do their related problems. The study of vehicle body structures provides an example of such refinement.

A major factor in holding down the cost of the current motor car whilst still improving its performance is increased structural efficiency. This brings, however, related problems of dynamic stiffness which are now manifesting themselves more and more strongly. It is perfectly possible for a body member to be structurally adequate in every way, except that it has dynamic properties which significantly contribute to vibration and interior noise levels.

Obviously, the body does not itself generate vibration but it does have properties which can amplify vibration and generate noise. Consequently, it is often argued that attacking noise problems by seeking to change the body is not very effective and that efforts should be concentrated on reducing vibration generation in the power train and suspension, while improving the isolation of the body from these sources. This has yet to be proved. The reduction of noise generation at source can be costly and the noise isolation of suspension and power-train is always in conflict with their location requirements.

No single approach can be completely correct but it has been demonstrated that the body can significantly affect noise characteristics. Many nominally identical prototype bodies have been found to have individualities which persist, even though all mechanical parts are changed. There are several possible explanations but the specific cause can only be determined with greater understanding of body vibrational behaviour. This in turn involves advancement of investigation techniques.

Having observed such individual variations, it is logical to suppose that changes to a body structure will affect its characteristics, and this has proved to be the case. To study these characteristics, use has been made of mobility methods, mobility being defined as the ratio velocity response/force input.

Even without use of mobility as a mathematical quantity, these methods yield and define the following important information:

The response to vibration of a particular input point.
The resonant frequencies of the structure.
The mode shapes and damping factors associated with the resonant frequencies.

For development work, such information provides useful further aid in tracking down, and solving, noise and vibration problems.

For design work, the information derived from various structures and tests should eventually enable us to:

Aim for body structures with low mobility levels and low amplitudes of vibration in the passenger compartment.

Avoid having potential input points at important antinodes.

Avoid mounting seats at important antinodes.

Avoid high body-response in passenger-sensitive locations.

The MS. of this paper was received at the Institution on 22nd April 1970 and accepted for publication on 7th May 1970.

* *Principal Research Engineer, Ford Motor Company Ltd, Research and Engineering Centre, Laindon, Essex.*

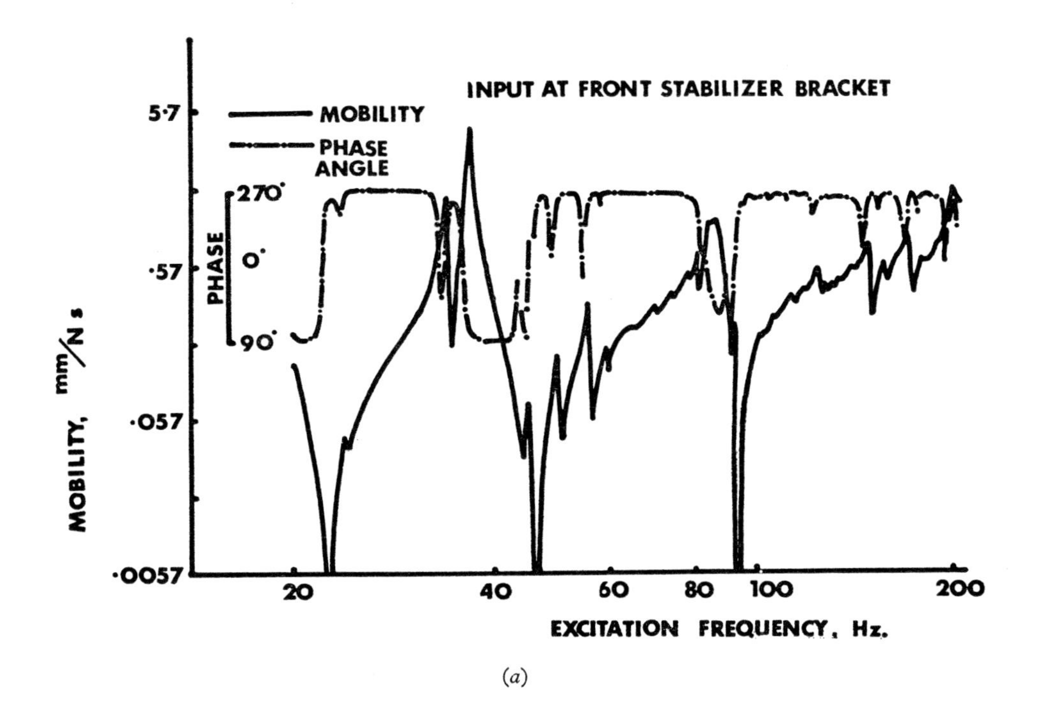

(*a*)

To summarize this introduction to mobility testing, it can be said that the body's response to vibration has a significant effect upon interior noise levels, and that changes to the body can be made which reduce these levels. What is lacking is a proper understanding of the relationships between body design and noise in service.

The purpose of this paper is to provide some of the answers and techniques which are necessary to achieve this understanding.

BACKGROUND

All the work described in this paper is aimed at a better understanding of the vehicle body's contribution to interior noise and vibration. In service conditions, the body is subject to complex, random and periodic, excitation at a number of input points. In an early attempt to isolate body response in a complete vehicle under service conditions, accelerometers were attached to the body and power spectral density (PSD) methods used to isolate periodic content. Cross PSD was employed to relate pairs of accelerometers to decide whether any specific body modes of vibration were present.

Although this provided valuable early experience in PSD analysis, the attempt was not very successful. The probable reason was that the road inputs to the vehicle were too large and swamped the body responses.

One way of improving this situation would have been by simulation methods, where greater control, and some simplification, of inputs can be achieved, for example, with a chassis dynamometer or electrohydraulic vibrator 'ride rig'. But even these would not have solved the

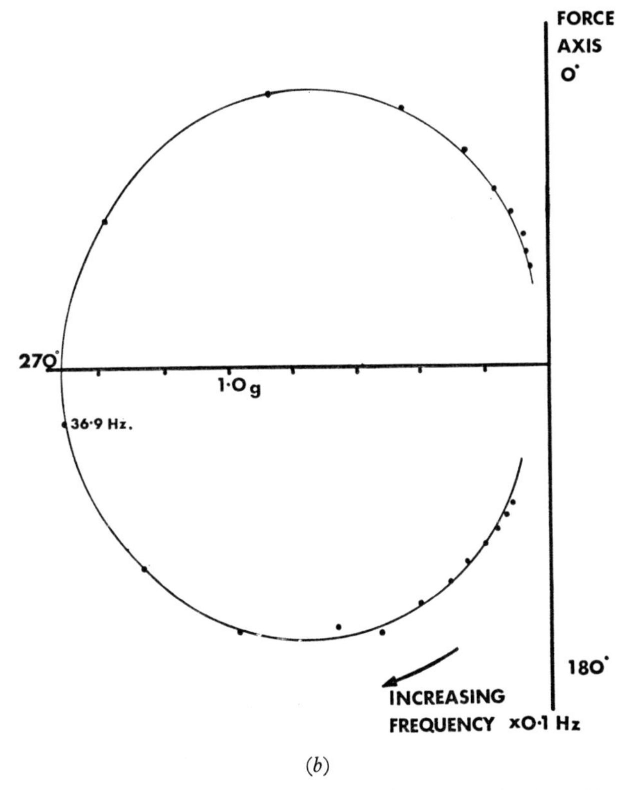

(*b*)

Fig. 14.1. Mobility and phase are plotted against excitation frequency (*a*) to decide the best input positions for the resonances which are of interest. The polar diagram (*b*) of input point response is then plotted and the mode shape (*c*) measured with a roving accelerometer

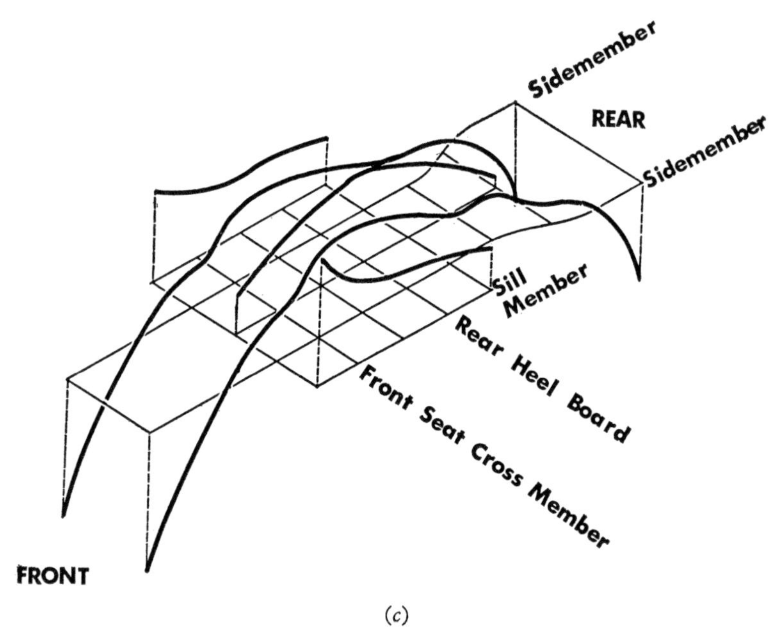

(*c*)
Fig. 14.1—*contd*

problem completely, because body-mode shapes derived by using real or simulated inputs under service conditions will always be impure because of the large non-resonant contributions. Reasonably pure modes are required for effective modification of the body.

The next stage of simplification was to use single-point sinusoidal excitation of the complete vehicle, but tests again proved this to be unsatisfactory and clear mode-shapes were very difficult to excite. It became necessary to attack the problem with a more fundamental approach by testing bare body-shells, which proved much easier to excite in definable resonant modes of vibration.

The justification for testing such a small part of the total vehicle system is that characteristics of body-response which result in high amplitudes in the shell are likely to do so in a complete vehicle, even though the frequencies and the relative importance of different modes will be altered by mass and damping changes.

The limitations these changes impose on the uses of body-shell results will be reduced by better understanding of relationships between the responses of the body shell and that of the complete vehicle; and between the response to pure sinusoidal excitation and that to the excitation experienced in service.

The first relationship is best investigated with the aid of multi-point excitation equipment of the type referred to later. The second is to be investigated by comparing multi-point excitation with chassis dynamometer or ride rig.

CURRENT TECHNIQUES

Limits of investigation

To bring the problem of defining a vehicle-body's response to vibration down to a feasible size, certain limitations have had to be accepted. Only vertical excitation has been used and, in general, only vertical response measured. The frequency range investigated is between 20 and 200 Hz for point mobilities, and is usually reduced to 20 to 60 Hz for the determination of mode shapes. Investigations have been concentrated on body shells without doors, bonnet, deck lid or glass. Bolt-on structural members, e.g. engine and gearbox cross-members, are normally fitted.

General method

The general method followed is to plot mobility against frequency at a number of input points (Fig. 14.1*a*). These groups indicate approximate resonant frequencies, including those at which mode shapes are to be determined. Having decided on the probable best input positions to excite each resonant frequency which is of interest, a polar diagram of input point response is plotted and a close estimate of natural frequency made (Fig. 14.1*b*). The mode shape (Fig. 14.1*c*) is then measured with a roving accelerometer, whilst this resonant frequency is held constant. For convenience in instrumentation the response parameter used is velocity for all mobility graphs and mode shapes, and acceleration for all polar graphs.

Selection of input points

The input points for which mobility frequency graphs are to be obtained are selected with regard to the following requirements. Points likely to have a very high mobility are avoided as the local response may mask much of

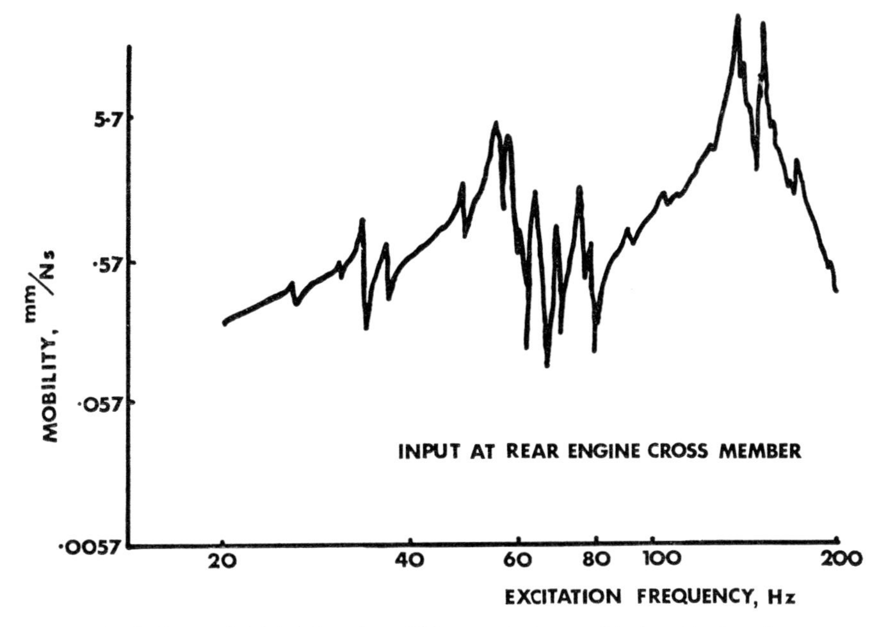

Fig. 14.2. Points with high local mobility are to be avoided as the local response can mask the structural response: compare this graph with Fig. 14.1*a* between 80 and 200 Hz

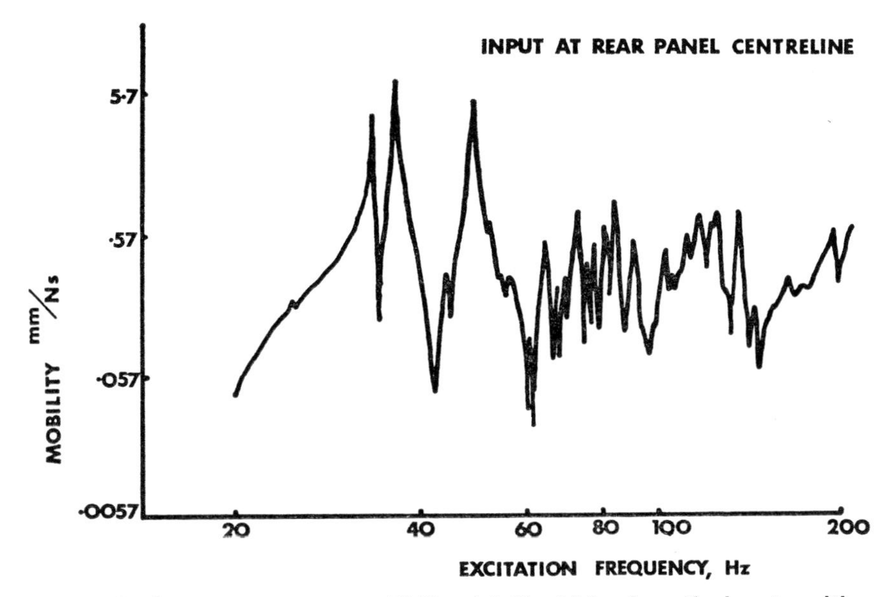

Fig. 14.3. Compare response at 50 Hz with Fig. 14.1*a*: here the input position is much further from the node at the front end, giving 10 times the amplitude

the structural response: compare Fig. 14.1*a* with Fig. 14.2 in the frequency range 80–200 Hz.

Enough points around the body must be investigated to avoid the possibility of missing a resonance through trying to excite it at a node. For example, the response at 50 Hz is much lower in Fig. 14.1*a* than in Fig. 14.3 because the former's input position is closer to the node at the front-end, which is apparent in the mode shape of Fig. 14.4*a*.

Of these points, some should be on the body centre-line and some offset, in order to improve the separation of beam and torsional modes at close frequencies, for example, the modes at 50·1, and 51·4 Hz shown in Fig. 14.4. The beam contributes only slightly in the torsional mode, and vice versa.

It is also normal to include all power-train and chassis mounting points as this may give an immediate guide to potential or actual problems.

Table 14.1. Peak mobility magnitudes for different input points

Frequency	Centre-line				Offset					
	Front panel	Engine rear cross-member	Rear panel	Drive shaft centre bearing	Front suspension front mounting	Front suspension rear mounting	Rear suspension front mounting	Rear suspension rear mounting	Rear damper mounting	Front jacking point
25										
26										
27										
28										
29										
30										
31										
32					3·4	2·5	2·4		0·9	2·3
33								1·5		
34	1·6	1·0	4·0	0·8	0·3				2·0	1·7
35						0·2	0·08			
36								1·1		
37	4·5	0·7	5·8	1·6			0·01		0·2	0·3
38					2·0	0·85				
39										
40										
41										
42										
43										
44										
45										
46										
47							0·2	0·23	0·35	
48					3·1					1·7
49		1·7			1·4	0·46				
50	0·2		5·0	2·3						
51										
52					0·57		0·20		0·14	
53						0·1		0·23		0·57
54										
55										
56	0·05									
57		5·1	0·65	5·2	0·5		0·42	0·32	0·17	1·1
58					0·22					
59	0·28	4·3	0·11	4·0			0·17		2·9	0·15
60							0·23	0·51		
61										
62										
63										
64										
65										

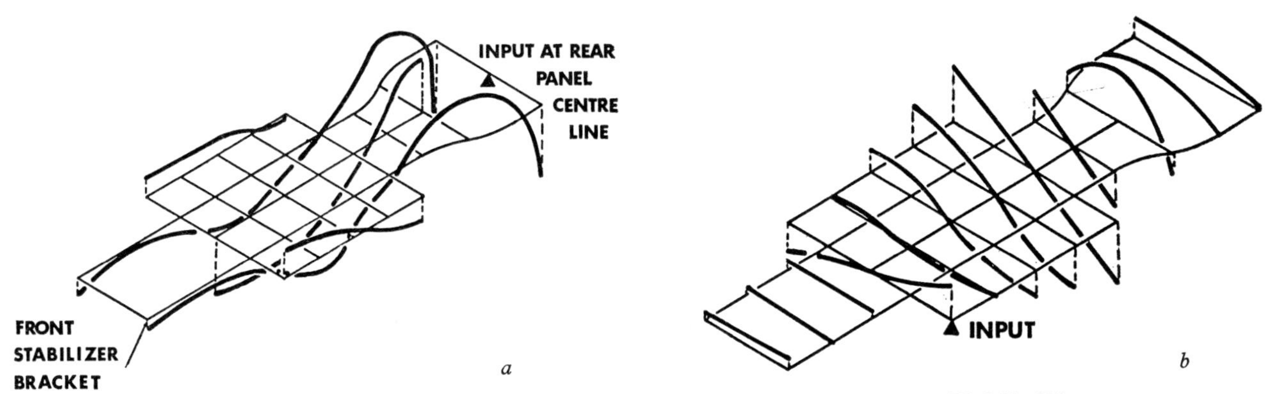

Fig. 14.4. Beam mode shape at 50·1 Hz (*a*) and torsional mode at 51·4 Hz (*b*)

The study of all the mobility graphs obtained in this way is often assisted by charting the magnitudes and frequencies of the peaks of each, as shown in Table 14.1. This rapidly highlights the outstanding resonances and indicates whether they are beam or torsional. It is usual to select the input point with the largest mobility for mode-shape investigation, unless this high mobility appears to be the result of local flexibility. Even then, for mode-shape investigation the first-chosen points should not be considered final. Plotting the mode-shapes obtained often indicates a better input position and a purer mode usually results if that point is used.

Determination of resonant frequencies

The determination of the exact natural frequency of a mode depends to some extent upon which resonance criteria have been chosen. In work concerned with relating structural vibration to noise and in comparing its shapes and magnitudes, a logical criterion is that of maximum response. This, however, does not take into account that, where there are resonances at close frequency intervals, the response at the natural frequency of one mode is affected by off-resonance contributions from other modes.

The methods of Kennedy and Pancu enable the effects of these modes to be separated and more accurate natural frequencies to be derived. In practice, the maximum frequency-spacing on a polar response graph is usually fairly close to the maximum amplitude; an example is shown in Fig. 14.5.

The use of resonant phase-angle (i.e. quadrature acceleration response) criteria would often yield significantly different frequencies, although correct placing of the exciter can, by reducing modal impurities, reduce this discrepancy.

Determination of mode-shape

The mode-shape of a particular resonance is obtained by holding the excitation constant at the resonant frequency and using a roving accelerometer to measure the response at a number of points on the body.

The accelerometer is mounted on a magnetic base and takes about 30 minutes to monitor 50 points. This results in a mass of data on magnitude and phase for all the points monitored.

There is always some phase scatter (Fig. 14.6), and, in order to derive a mode-shape which is true at some instant, it is necessary to correct all magnitudes for phase angle. The angle chosen as basis is not necessarily that at the input point. A simple method is used to obtain the phase angle for which the sum of all the monophase components is a maximum. The mode-shape is then plotted by using these components. Monophase means phase axis of ϕ or $\phi+180$ degrees).

The amount of phase scatter is an indication of modal

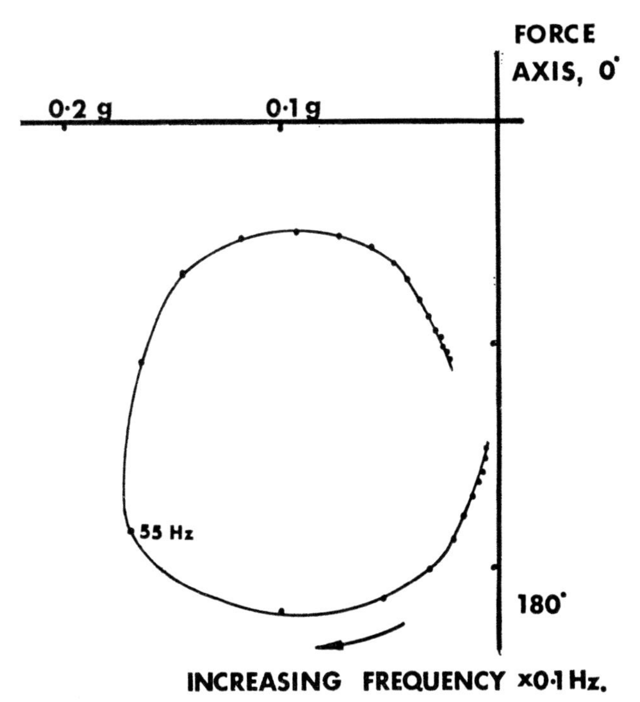

Fig. 14.5. Acceleration response at input: maximum frequency spacing is fairly close to maximum amplitude

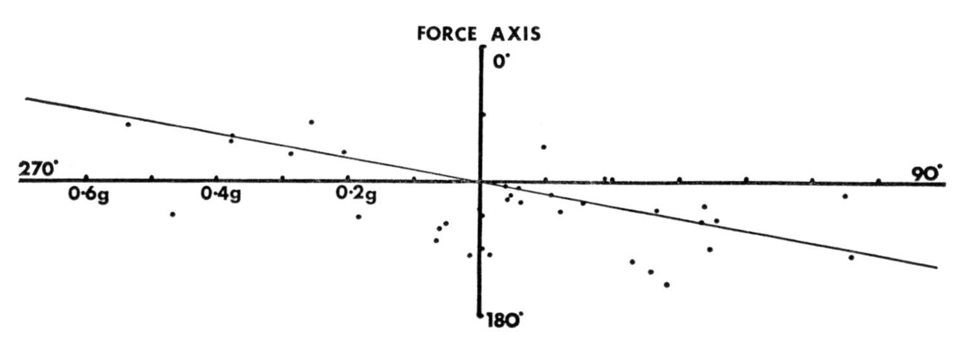

Fig. 14.6. Acceleration response of monitored points on body: there is always some scatter about the general phase angle

purity and experimental accuracy. It is sometimes useful to calculate the sum of the quadrature components to the chosen phase angle.

INSTRUMENTATION

The instrumentation used is based on a commercially available system, the main points of which are shown in Fig. 14.7.

Excitation is provided by a single electromagnetic vibrator which is suspended vertically from the input point. Reaction for the vibratory force input is provided by the inertia of the vibrator casing and coils. It is not considered that the dead-weight exerted by the vibrator has any significant effect on dynamic response.

The system monitors force and acceleration signals from transducers in an impedance head mounted between the exciter and the body. Both force and acceleration signals are taken through tracking filters tuned by the control oscillator. Servo loops are used to multiply each signal up to the level required for optimum working of the filters.

The magnitude of each signal is derived from the gain of its respective servo loop. The conversion of acceleration amplitude to velocity amplitude, and the division of velocity by force, are accomplished logarithmically and the output plotted against frequency. Measurement of the phase angle between force and velocity is obtained from high-frequency modulated signals generated by the tracking filters.

A compressor circuit, controlled by the force-transducer signal, is used to maintain constant force input for graphs of mobility against frequency. The effects of mass between the impedance head and the body are cancelled out by a system which vectorially subtracts a signal proportional to the acceleration from the force signal. The constant of proportionality is set by adjusting a potentiometer so that the force signal is brought to zero when the exciter is running with all the attachment brackets fitted, but disconnected from the body. This form of mass cancellation works on the response side of the system and ensures true values of mobility but it is worth noting that the compressor circuit for control of force works on the actual transducer signal. This means that the mobility/frequency is not, strictly, plotted at constant force. In order to interpret the graphs the force/velocity relation must be assumed to be linear. For the body shell, this is a perfectly reasonable assumption.

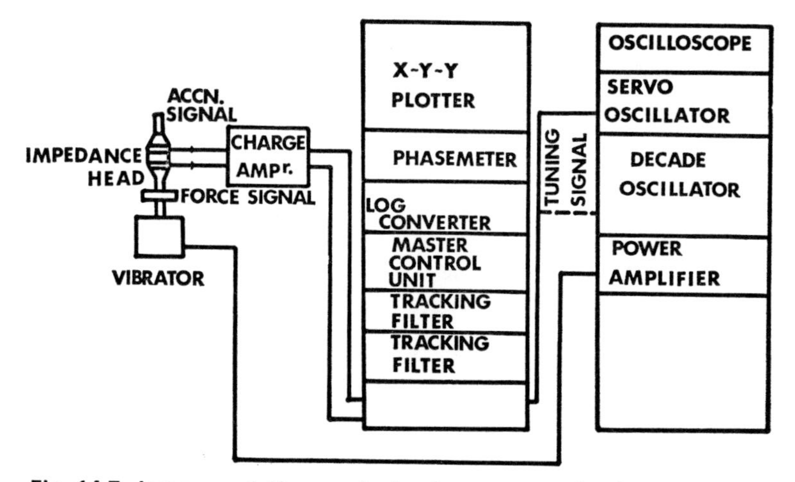

Fig. 14.7. Instrumentation: a single electromagnetic vibrator provides the external excitation and the system monitors force and acceleration

MODIFICATION AND CHANGES TO THE BODY

Scatter

Before the significance of a change in response can be evaluated, it is necessary to know the amount of scatter which may exist between nominally similar unmodified bodies. Some attention has been paid to this by testing two nominally identical prototypes; and three nominally identical production bodies.

The two prototypes had been assembled by hand, and contained many hand-beaten parts. They showed noticeable differences in characteristics above 50 Hz (Fig. 14.8) and also some differences in mode-shapes (Fig. 14.9).

Of the production bodies, two were manufactured consecutively and a third was produced 12 weeks later. The important findings from these tests are as follows:

(1) The three bodies had mobility characteristics immediately recognizable as very similar. The same resonances occurred with fairly small differences in frequency and magnitude (Fig. 14.10).

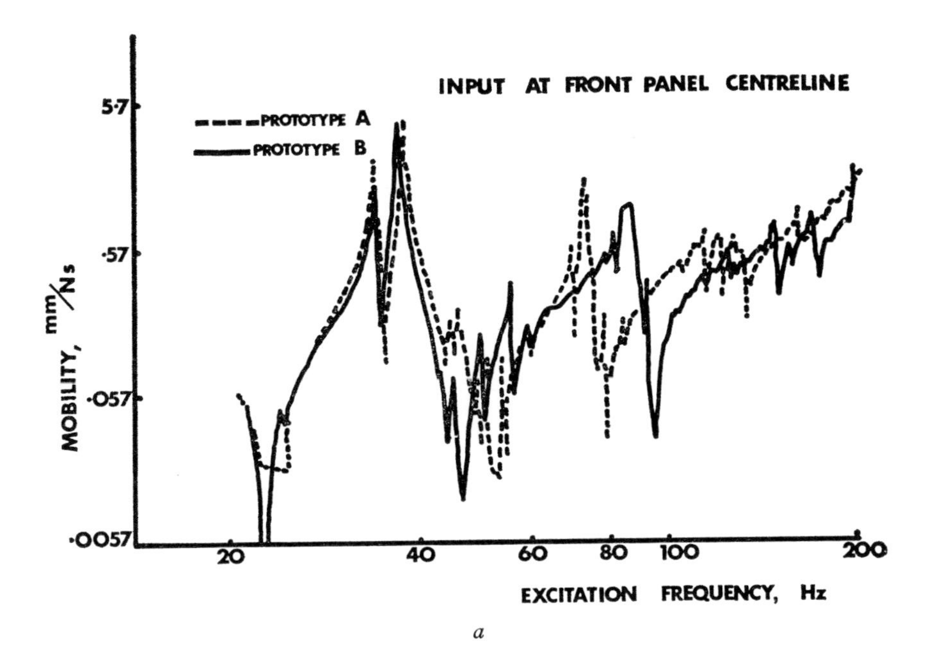

a

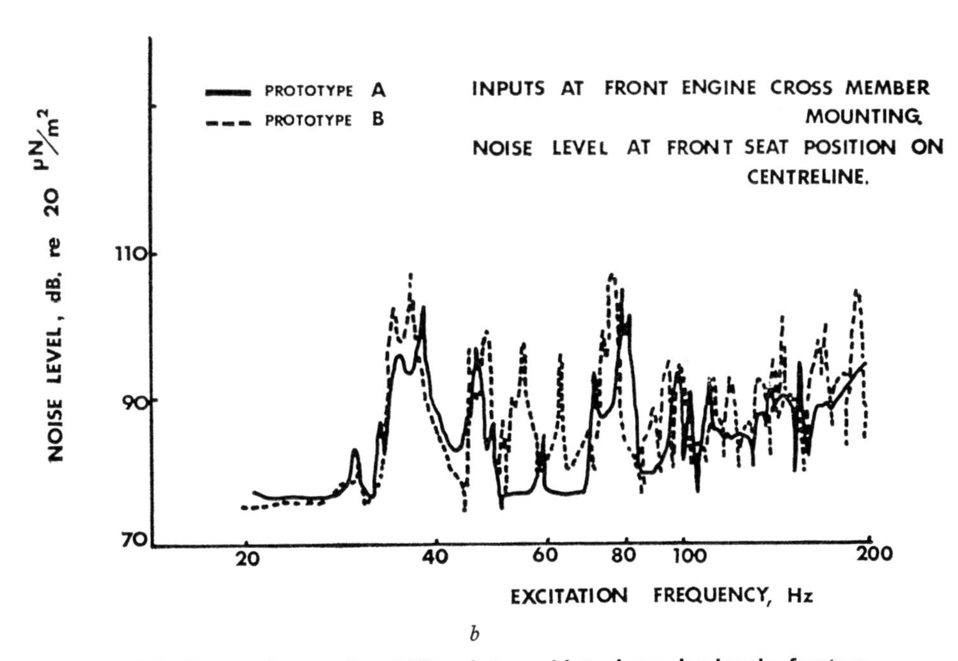

b

Fig. 14.8. Comparisons of mobility plots and interior noise levels, for two nominally identical prototypes

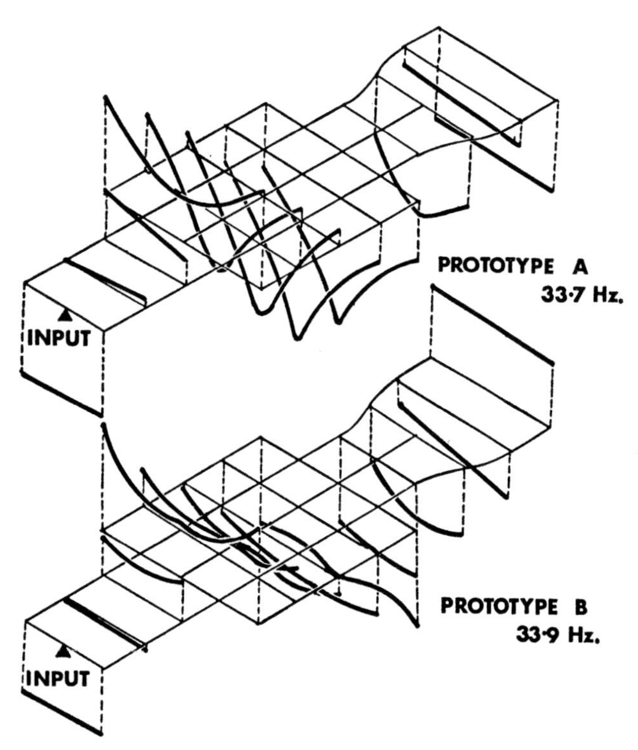

Fig. 14.9. Comparison of mode-shapes of nominally identical bodies

(2) The differences between the mobility graphs for the three different bodies are slightly greater than the differences which can exist between opposite sides of the same body. Table 14.2 illustrates this by showing the ratios of resonant response between sides and between bodies. There is good agreement between damping factors and response for the three which extends to noise levels; body 33 had least damping, highest response and greatest noise, Fig. 14.11 compares noise levels of bodies 33 and 34.

(3) The mode-shapes associated with corresponding resonances are very similar, although differences of magnitude are more obvious because they are plotted on a linear basis, whereas the mobility plots are logarithmic. Fig. 14.12 shows two of the more dissimilar cases.

Modifications

To alter a vehicle body's response changes can be made in damping and stiffness. Since the aim is to design shells which will avoid body response problems, efforts have largely been directed at changes in stiffness.

A number of modifications were made by adding members to a body-shell, and the effects of these on both noise and velocity response were measured. It was found possible, but not easy, to materially alter velocity response

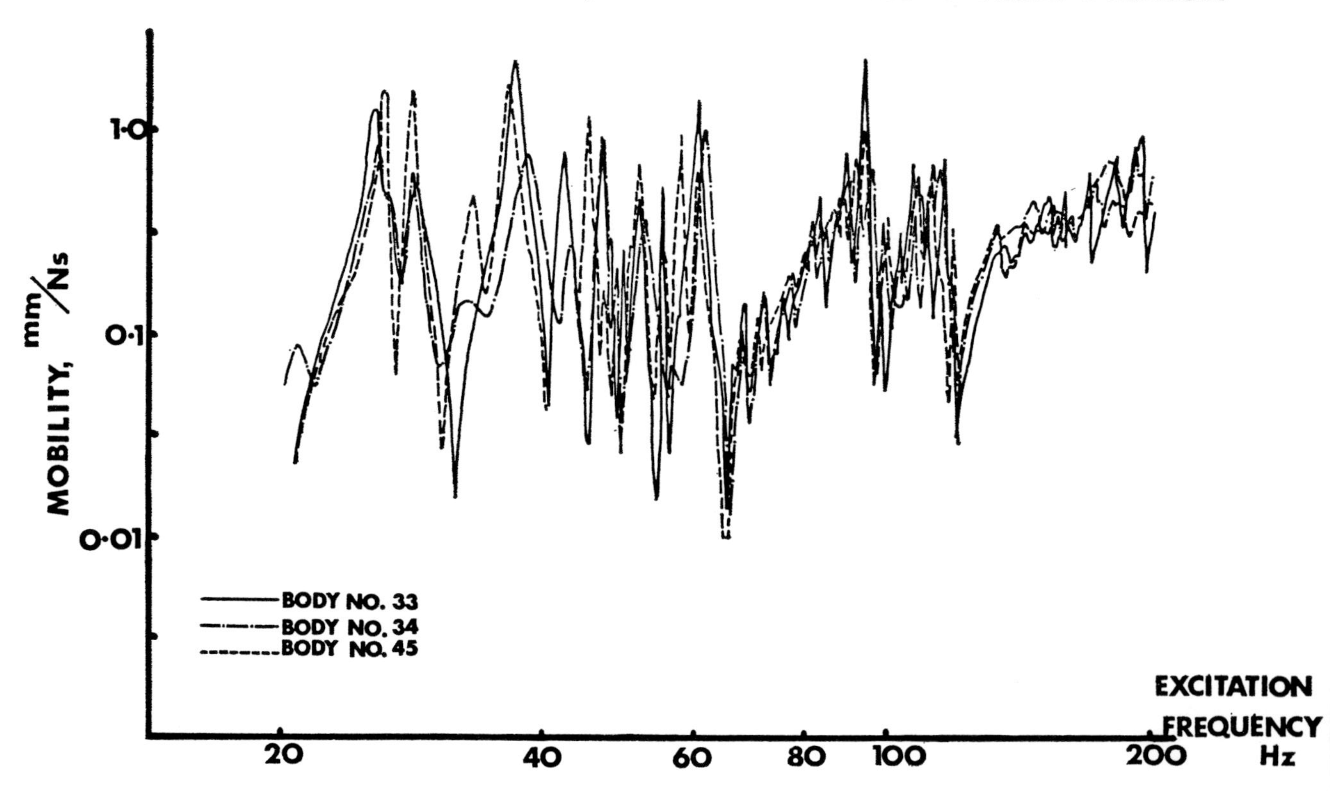

Fig. 14.10. The mobility plots of three production bodies, one of which was produced 12 weeks after the other two: differences are small

Table 14.2. To show ratios of maximum amplitudes of mobility response for similar resonant modes, the respective damping factors, and to show deviation in resonant frequencies

Resonant frequency, Hz	Mode function	Input at right-hand rear spring, rear shackle			Damping factor, ($c/c_c \times 10^{-3}$)			Input position, body 34		Input position, body 45	
		Body 34	Body 45	Body 33	Body 34	Body 45	Body 33	Rear spring rear shackle r.h.	Rear spring rear shackle l.h.	Rear spring rear shackle r.h.	Rear spring rear shackle l.h.
25·7	Torsion	1		2·86	7·35	6·45	5·7	1	1·28		
26·3			1·41							1	0·71
27·6	Torsion				8·15	5	7·75				1·34
28·5		1	2·52	1·49				1	1·14	1	
36	Bend		2·08		18	10·5	9			1	1·32
38·1		1		2·7				1	1·06		
42·7	Bend	1	1·57	1·96	7	9·5	9·5	1	1·18	1	1·33
45·5	Torsion		1·34		4·8	4·65	3·95			1	0·55
46·7		1		2·86				1	0·55		
51·3	Torsion				11·5	3·65	5·9				0·69
52		1	2·26	1·61				1	0·47	1	
60·8	Bend		0·65	1·37	5·6	4·7	4·9			1	1·36
61·4		1						1	0·91		

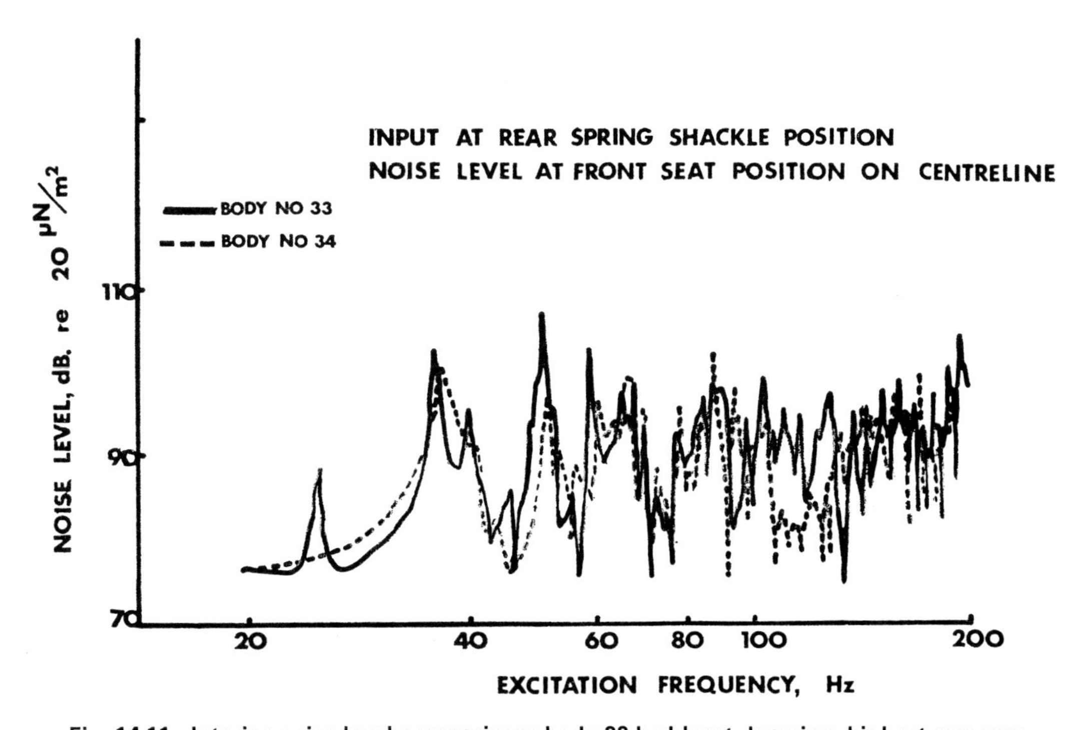

Fig. 14.11. Interior noise level comparison: body 33 had least damping, highest response and also greatest noise

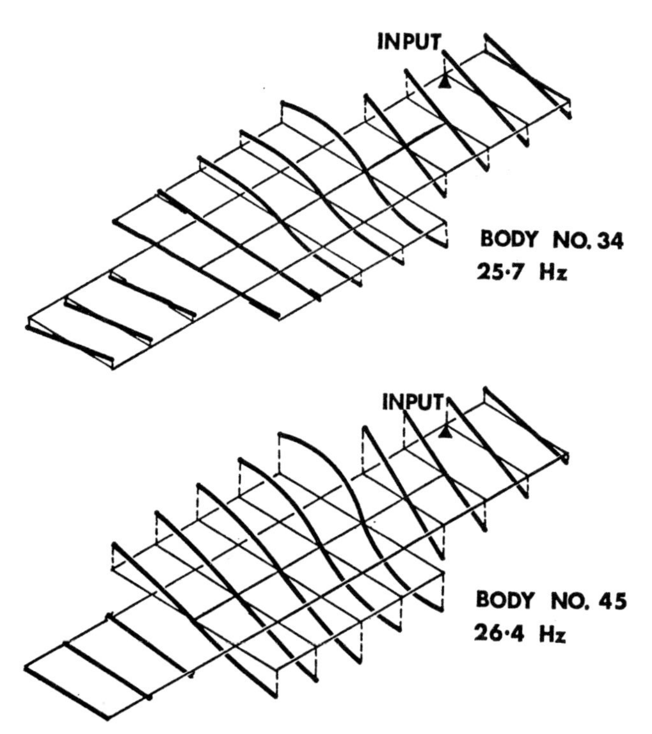

Fig. 14.12. Mode-shapes at about 26 Hz of two of the more dissimilar bodies: the differences are still small

whilst the frequencies of resonances were usually only slightly altered. Those modifications which seemed to produce the largest reductions in passenger-compartment velocity-response did not necessarily produce the greatest reduction in interior noise level.

For example, the added members shown in Fig. 14.13 produced the change in noise response shown in Fig. 14.14. The change in mode-shape at 35·7 Hz is shown in Fig. 14.15. The reduction in noise level around this frequency was greatest with the changes shown, and other changes to the body which produced even larger reductions in body response did not further reduce noise levels.

It is, of course, often found that reduction of response at one frequency may be offset by increased response at another, and it becomes necessary to have information on actual problem frequencies, or potential input frequencies, in order to assess the real improvement due to a modification.

For the same reason, the true effect of any modification can only be established if pure modes of vibration can be set up, so that the different effects of closely coupled modes can be separated.

An interesting exercise which, whilst not strictly a modification, certainly alters the response of the body, is the addition of windscreen and backlight glasses. Fig. 14.16 shows the difference in mobility/frequency graphs and noise levels. The change appears to be primarily in stiffness because, although the resonant frequencies differ, the levels of damping in the two cases are similar.

RELATIONSHIPS BETWEEN INTERIOR NOISE AND BODY MOTION

There is usually a basic correlation between noise and mobility which makes it logical to look for a link between the two, Fig. 14.17, especially as (to take the simplest, theoretical case) the sound-pressure generated by a flat piston is proportional to the piston velocity.

Test work on one model showed that, at most of the lower resonant frequencies by far the highest amplitudes occurred in the rear floor area. This appeared to be due to a lack of transverse dynamic stiffness and the modifications referred to in the previous section were aimed at reducing motion in this area.

The results indicated that a more accurate method of relating noise to body-motion was required. A test was carried out to investigate the volume-velocity/sound-

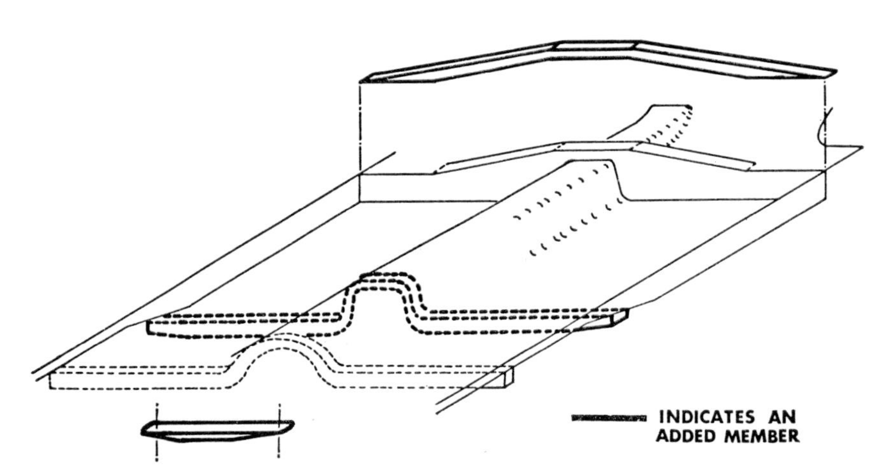

Fig. 14.13. The additional members shown produced the changes in noise response shown in Fig. 14.14: velocity response could be altered materially

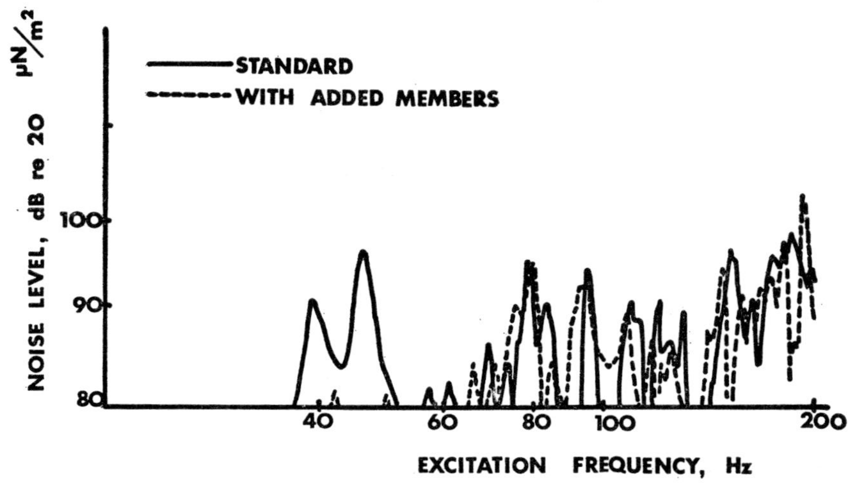

Fig. 14.14. Noise reduction at front seat position due to added members shown in Fig. 14.13: changes in resonant frequencies are small

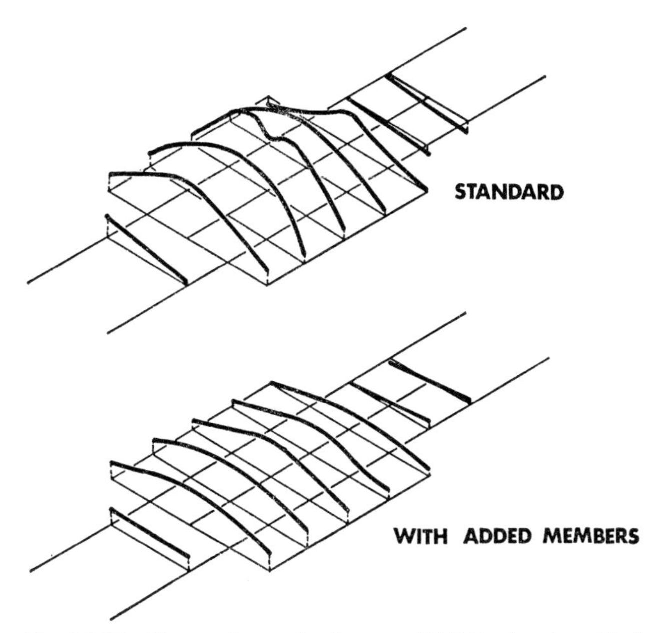

Fig. 14.15. Change in mode shape at 35·7 Hz due to added members: further changes which reduced body response did not further reduce noise

pressure relationship demonstrated by J. W. Dunn at the Motor Industry Research Association (1)*.

Volume-velocity

Volume-velocity is the net rate of change of interior volume of the body, caused by the movement of the noise-generating surfaces. Fig. 14.18 shows the relationship between the sound-pressures inside the passenger compartment and the corresponding volume-velocity of the floor and roof for a number of resonant modes. For each resonance, a range of sound pressures is given, this being the maximum and minimum levels found over four measurement points at passenger-ear positions. The straight line represents the simple theoretical relationship, its position depending on generation (piston) area.

The actual relationship is not as clear as one would hope but this may be for a number of reasons. The noise measurements may suffer from inaccuracies but the resonant noise levels inside the shell are not likely to be more than 2 dB in error. The actual measurements used by us are slightly different from Dunn's (1). The sound-pressures shown are those obtained over the front and rear-seat passengers' inner and outer ear positions. Volume-velocity levels are obtained from mode-shapes of total, rather than quadrature, response; and the volume velocities of floor and roof are calculated separately and added vectorially.

Other force, velocity and noise relationships

The Dunn relationship needs further refinement before it can be used for predictions. However, other tests (2)–(4) carried out to investigate force, velocity and noise relationships indicate that it is a useful approach. The main conclusions derived from these tests are as follows:

(1) The assumption of force/velocity linearity is reasonable. Fig. 14.19 shows input mobility for different force inputs. The graphs virtually coincide, except for a slight reduction with increased input, in both mobility response and resonant frequencies.

(2) For a given resonant mode, the noise generated is simply related to the force-input. Fig. 14.19 also shows noise responses for different inputs and Fig. 14.20 shows the force/noise relationship for certain frequencies.

It follows that, at a given resonance, the sound-pressure is closely related to velocity response. The remaining problem is the accurate determination of that function of velocity-response which dictates the noise-levels corresponding to different resonant modes.

* *References are given in Appendix 14.1.*

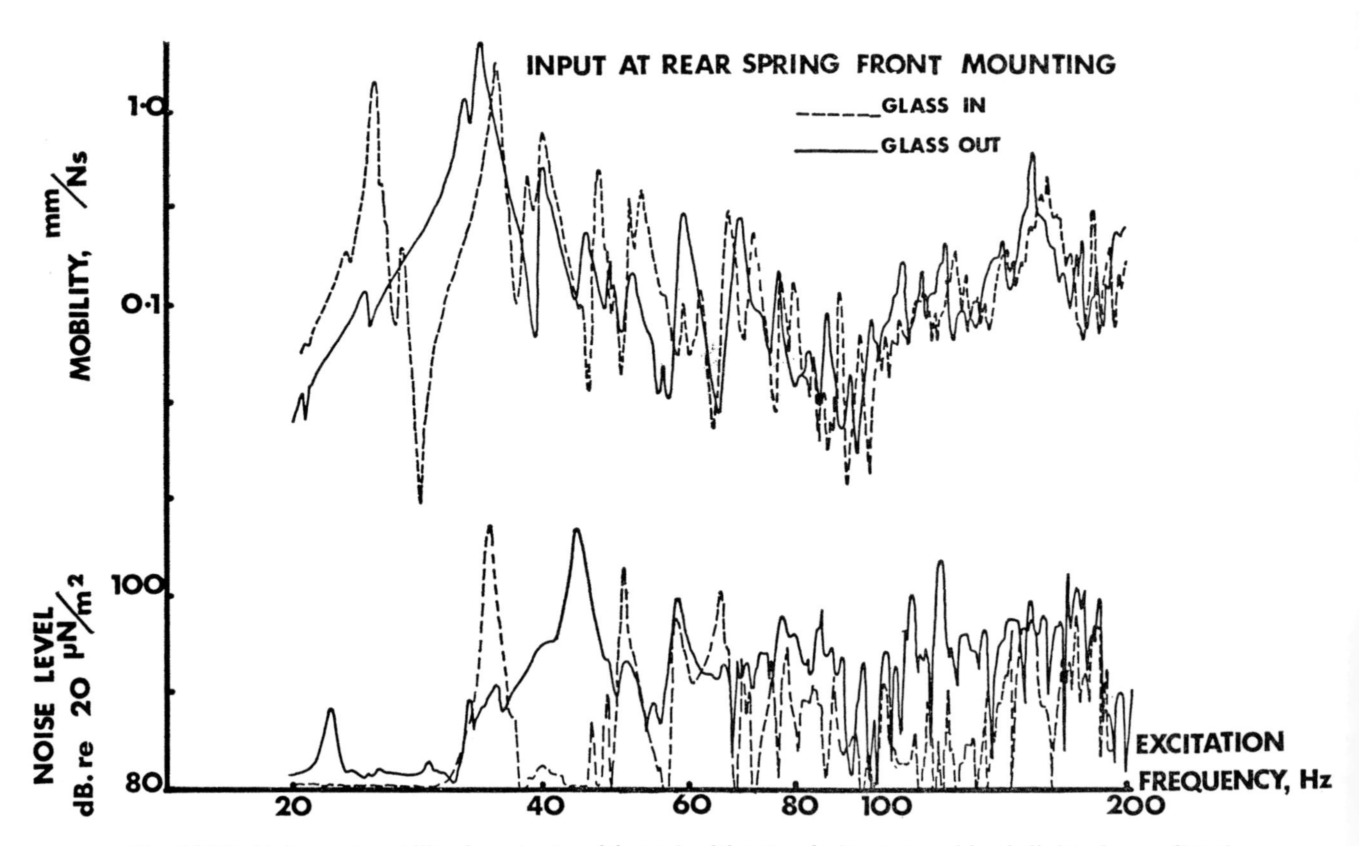

Fig. 14.16. Noise and mobility for a body with, and without, windscreen and back-light glasses fitted

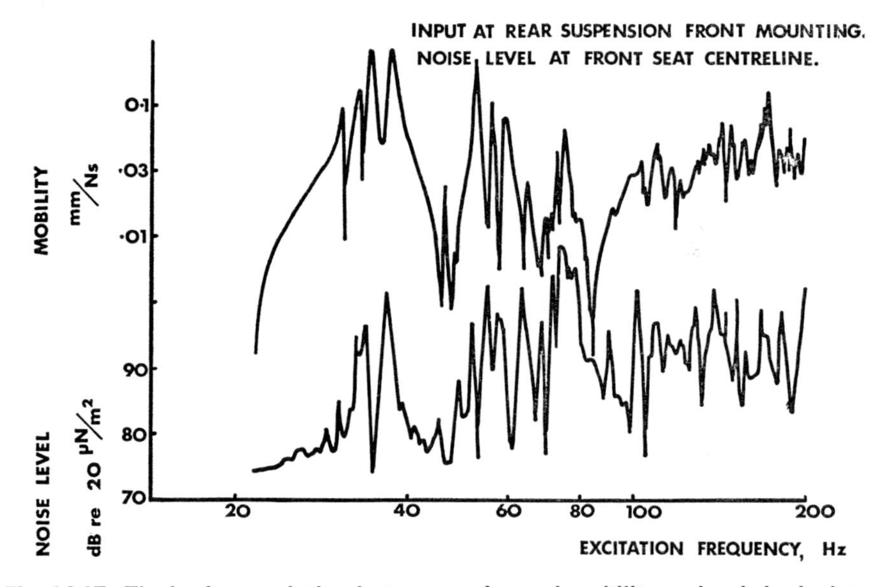

Fig. 14.17. The basic correlation between noise and mobility makes it logical to look for a link

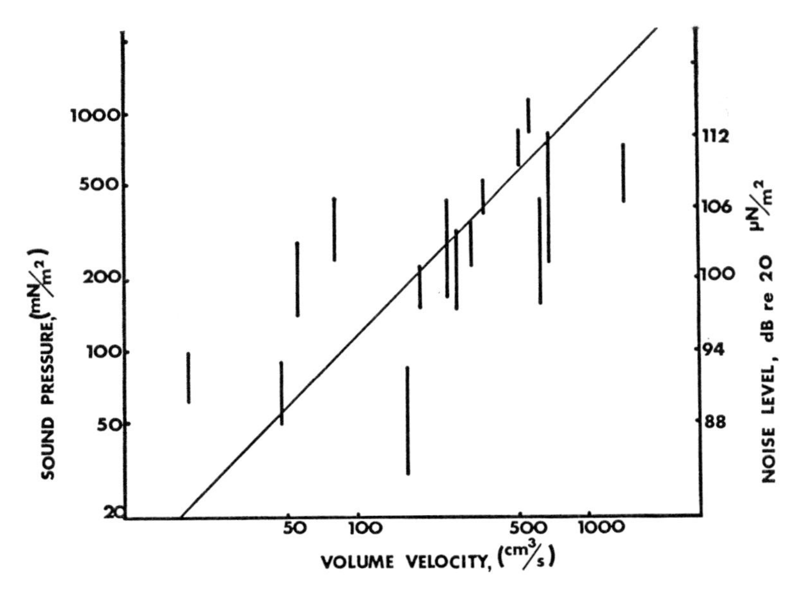

Fig. 14.18. Relationship between sound-pressure inside the passenger compartment and the corresponding volume-velocity of the floor and roof

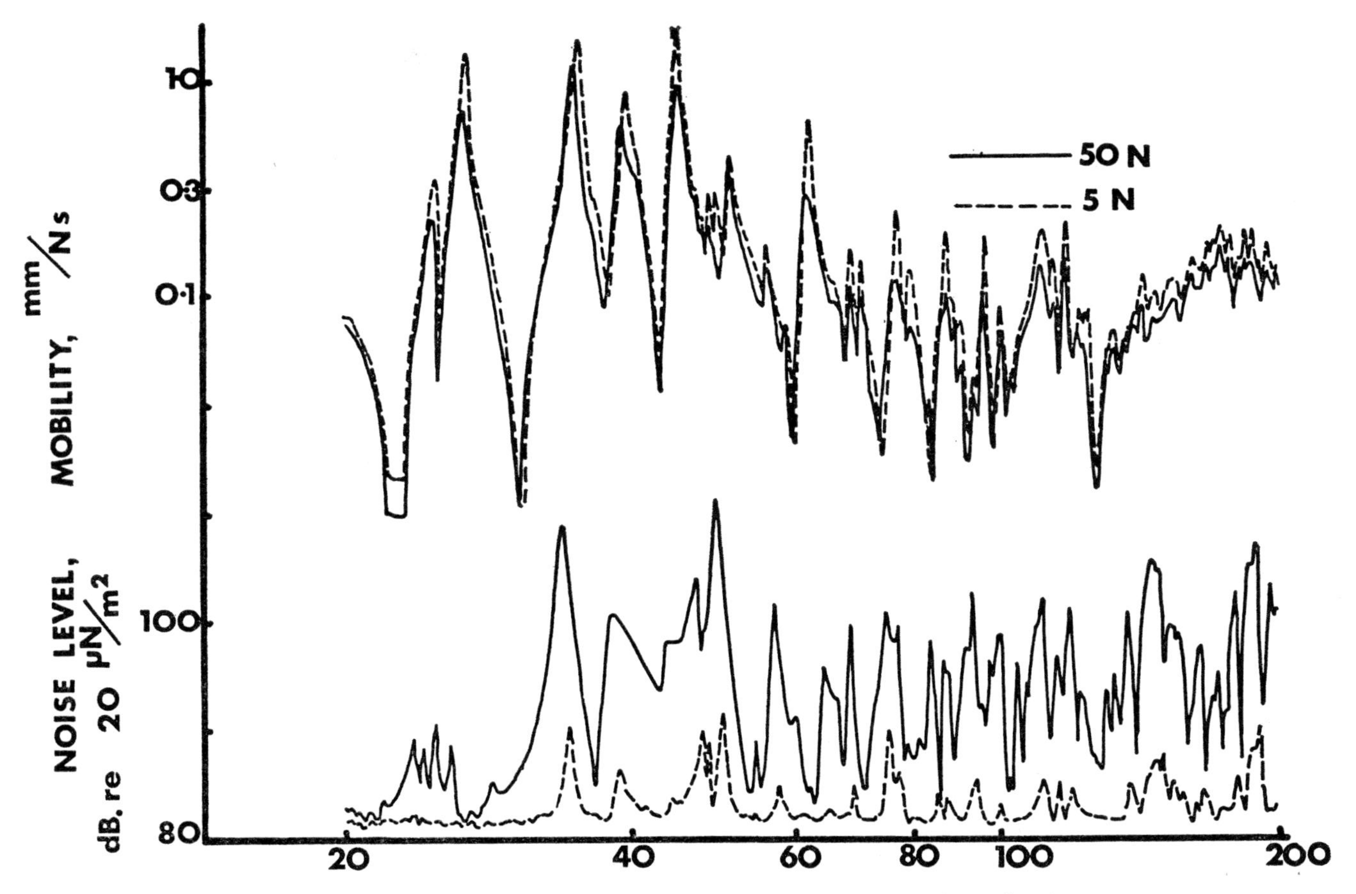

Fig. 14.19. Mobility and noise response at different force-input levels

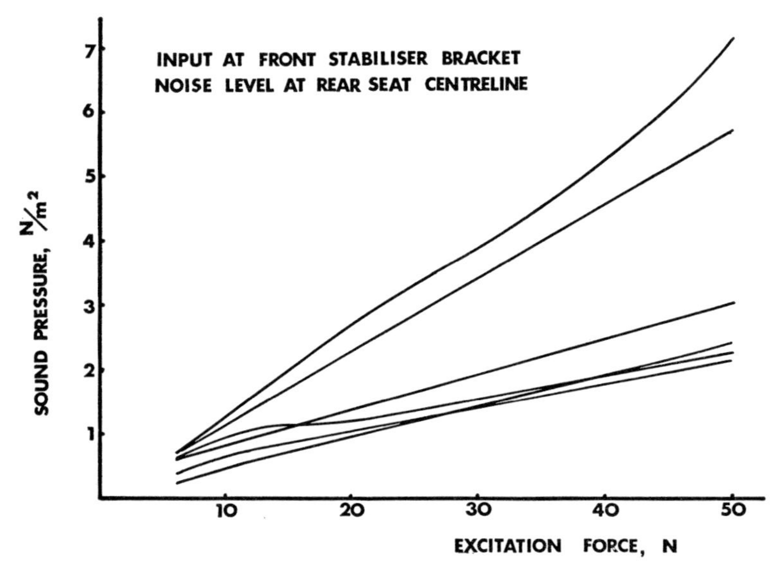

Fig. 14.20. For a given resonant mode, the noise generated is a simple function of excitation force

LIMITATIONS

Interaction of modes

The major limitation of any single-point excitation system is the frequent impossibility of studying a particular resonant condition without interference from off-resonant contributions by other modes.

A structure may have two resonances only a small interval apart. The response of such a system is shown by the solid line in Fig. 14.21. If either of these resonances existed independently, their responses would be as shown by the dotted lines: when both interact, the total response at one resonant frequency includes an off-resonant contribution from the other.

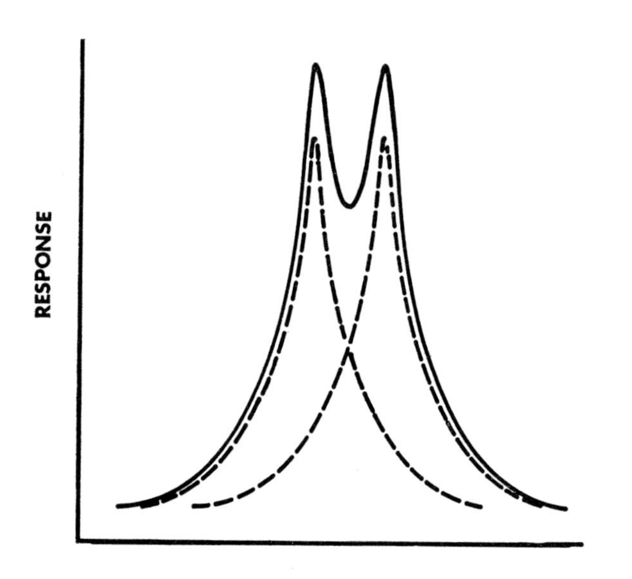

Fig. 14.21. The effect of two closely spaced resonance peaks on each other is shown in the full line: if either were present alone, one of the dotted graphs would apply

Since the effect of damping on resonance is to lower the maximum and widen the base of the response peak, it is apparent that both damping and the frequency interval between resonances affect the amount of off-resonant contributions. (This is referred to as 'coupling'.)

The phase effects of this interaction must also be considered. Each resonance would by itself have peak velocity-response in line with the force input but the off-resonant contributions of each mode to the other clearly are not in line. The resultant peak responses of both modes together are therefore affected in phase, as well as magnitude.

The effects of this can be largely accounted for by the methods of Kennedy and Pancu (**5**). The polar diagram of Fig. 14.5 is a case where a large off-resonant response can be separated out, but to obtain a pure mode-shape this would have to be done for every mode measuring point.

It is also possible to find resonances which do not yield circular polar response diagrams (**6**). Fig. 14.22 shows one in which two separate circles could be fitted, although there is only one point of maximum frequency spacing. One effect of this is to make the calculation of damping less accurate. It is also apparent in Fig. 14.22 that the resonant frequency derived from frequency spacing is not on the 270 degree resonant axis.

These effects become even more significant as the structure becomes more complex. The response graph of Fig. 14.23 was taken from a complete vehicle and the corresponding phase scatter for points about the underbody is shown in Fig. 14.24, which may be compared with the scatter on the body-shell only, shown in Fig. 14.6.

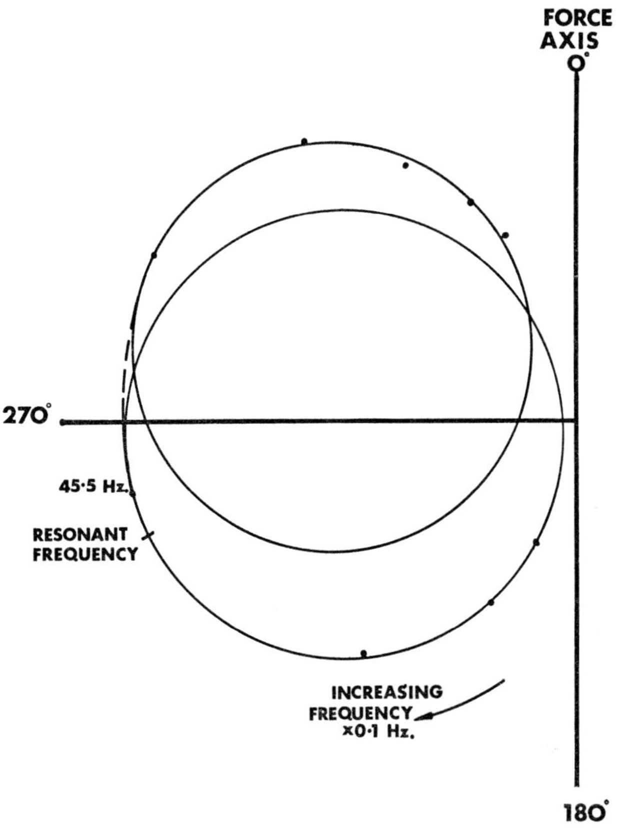

Fig. 14.22. In some resonance cases two separate circles could be fitted although there is only one point of maximum frequency spacing

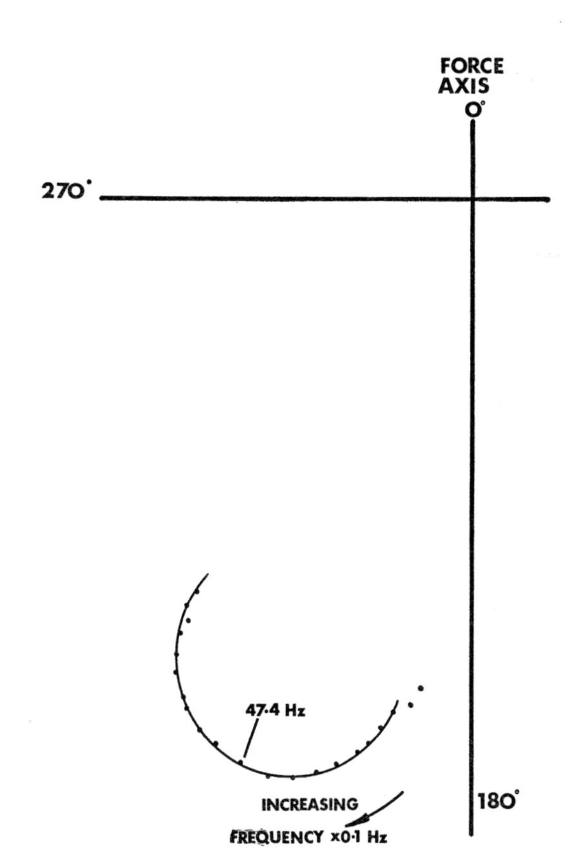

Fig. 14.23. Acceleration response of a complete vehicle

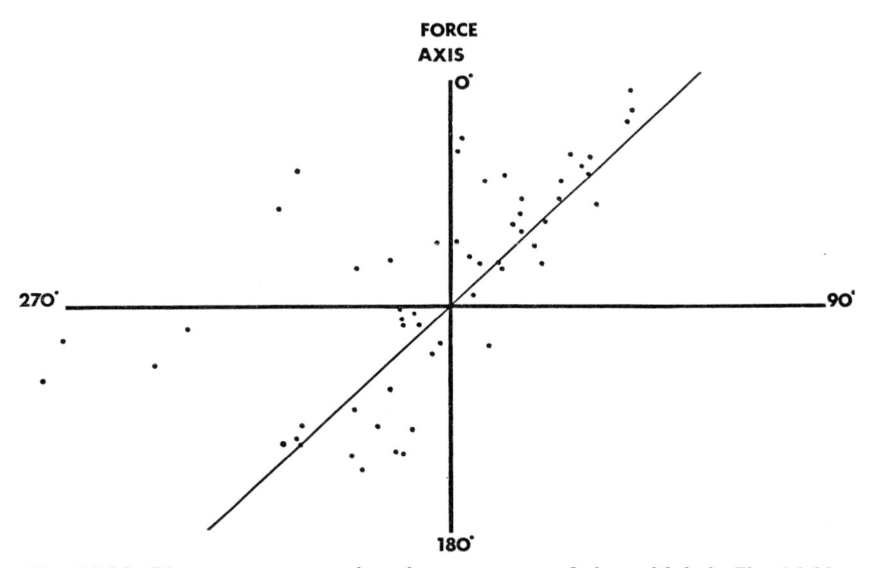

Fig. 14.24. Phase scatter acceleration response of the vehicle in Fig. 14.23, taken at various points: compare with Fig. 14.6

It is better to improve these conditions by greater control of input conditions than by further analysis of response. This will be discussed below.

Complexity

A rather different sort of limitation to the work carried out so far is the large volume of data required to define a body's behaviour. One cannot easily define the mobility of the structure as a whole, one has to define the mobility of every point as a function of frequency and also to define the input point. Similarly, a noise measurement depends upon position of measurement, frequency, and excitation input point.

The ultimate way of storing such data would be with the aid of a mathematical model but the problem of constructing such a model based on test results is probably as great as that of a mathematical idealization designed to predict the experimental results.

It must also be emphasized that mobility investigation can never be a solution by itself. Mobility is merely the relationship between input and response and only when two of these three are known can the third be predicted.

MULTI-POINT EXCITATION

In the context of this paper, multi-point excitation is a way of obtaining a pure response by appropriate control of inputs.

Modal impurities are caused by off-resonant responses altering the magnitude and phase of the response being monitored, so that all points on the structure are not in a monophase condition. Conversely if it is possible to excite the structure without forcing it so that it is entirely

Fig. 14.25. The MAMA apparatus

monophase along the correct resonant axis, then the structure must be in pure resonance.

This situation can be obtained by arranging inputs and measuring response, at several points. Input forces and frequency are then arranged so that the off-resonant contributions are suppressed.

This degree of control is achieved in the Royal Aircraft Establishment's multi-point system known as MAMA (Manual–Automatic Multi-point Apparatus) (Fig. 14.25) (7). The frequency of all the exciters is the same and is controlled by a phase/frequency servo-loop which compares the phase between force and response at one 'master' input position, with a demanded resonant phase-angle. The error signal is then used to alter the frequency so as to 'home' on a resonance. While the frequency and master-vibrator remain locked on to the resonant condition, it is possible to adjust the force levels of the other vibrators until all the response measurements are monophase; this is the true resonant condition.

Most of the advantages of MAMA lie in overcoming the limitations of single-point excitation mentioned earlier. In practice, there may well be additional benefits. It may prove possible to excite, to an acceptable level of purity, all the resonances of interest with only one layout of input positions. This will reduce time taken up in the normal iterative procedure for finding the best input points. The excitation of pure modes will make it unnecessary to process response data to find a general phase-angle and there will be no contradiction between the various resonance criteria.

Initial exploratory testing with MAMA of a body of known properties (8) has already demonstrated the ability of the system to lock on to all the resonances previously found, and to hold demanded phase angles near the resonant axis.

CONCLUSIONS

We have outlined the single-point excitation technique used by Ford Motor Company as one of the methods for investigating the vibrational behaviour of vehicles.

This technique is adequate to define the basic behaviour of bare body shells. In addition to its use as a development tool it has shown the sort of variations which can be expected between nominally similar bodies; and that modifications can significantly reduce vibration and noise response.

There are limitations to this procedure in accurately assessing the effects of body modifications, and in correlating body shells with complete vehicles. Many of these can be overcome by improving the excitation system rather than the analysis and the case is presented for a multi-point excitation system which has great potential in this work.

ACKNOWLEDGEMENTS

The author wishes to thank the Ford Motor Company Limited for permission to publish this paper, and to recognize the invaluable assistance of his colleagues.

APPENDIX 14.1

REFERENCES

(1) DUNN, J. W. and ASPINALL, D. T. 'A study of the vibration and acoustic characteristics of saloon car bodies', *MIRA Report* 1969/15.

(2) CRAGGS, A. 'The low frequency response of car bodies', *MIRA Report* 1965/14.

(3) TRAIL-NASH, R. W. 'On the excitation of pure natural modes in aircraft resonance testing', *J. Aero/Space Sciences*, December **25**, 1958; pp. 775–778.

(4) BISHOP, R. E. D. and GLADWELL, G. M. L. 'An investigation into the theory of resonance testing', *Aeronautical Research Council A.I. Report ARC* **22**, 381, O. 1596, September 1960.

(5) KENNEDY, C. C. and PANCU, C. D. P. 'Use of vectors in vibration measurement and analysis', *J. Aero/Space Sciences*, 1947 Volume **14** page 603.

(6) PENDERED, J. W. and BISHOP, R. E. D. 'A critical introduction to some industrial resonance testing techniques', *Journal Mechanical Engineering Science* Volume **5** No. 4 1963.

(7) TAYLOR, G. A., GAUKROGER, D. R. and SKINGLE, C. W. 'M.A.M.A.—A semi automatic technique for exciting the principal modes of vibration of complex structures', *Royal Aircraft Establishment Report* TR 67211 August 1967.

(8) GAUKROGER, D. R. and HAWKINS, F. J. 'Resonance tests on a motor car body', *Royal Aircraft Establishment Technical Report* 66022 January 1966.

Paper 15

SOUND DAMPING CONTROL OF AUTOMOBILE BODY STRUCTURES

A. M. Chappuis*

The acoustic comfort of cars is considered on the bases of measurements and subjective judgement. The two causes of discomfort, resonance and excessive sound level, are discussed separately, with special emphasis on resonances of the car body. The sources and transmission of noise are considered and different techniques of noise reduction examined. Sound insulation is treated in detail.

INTRODUCTION

IT IS INTERESTING to consider some factors which have caused the rapid escalation of noise problems in motor cars. These factors belong to two categories, those concerned with subjective comfort and those arising from technical progress.

Acoustic comfort is only one aspect of general comfort and, as considerable advances have been made in the other fields, such as ventilation and heating, seating, etc., there has been demand for similar progress in acoustics. In addition, noise has been proclaimed one of the plagues of modern civilization and the public has became highly sensitive to it. As a result, customers have become more exacting on questions of car noise.

On the technical side, the factors are numerous but one is particularly important. This is the tendency to be found in nearly all means of transport for the power of the engine to be increased whereas the weight of the vehicle is reduced in order to achieve better performances. Studies on ships have shown that the importance of noise and vibration problems increases with the square of the ratio of power to weight. Although it is more difficult to verify, and less pronounced, in motor cars, this empirical rule remains important.

The other technical factors include the increase in speed and power/weight ratio of the engine itself, the introduction of overhead camshafts on touring cars and the more elaborate tyre patterns. All of them usually increase the noise levels.

The MS. of this paper was received at the Institution on 22nd June 1970 and accepted for publication on 30th June 1970.

* *Research and Development Engineer, Interkeller AG, 8052, Zürich, Schärenmoostr., 105, Switzerland.*

The monocoque body, where rigidity is replaced by flexibility, has rendered the problem of mounting vibrating parts much more difficult. In fact, it is evident that practically all the efforts made to improve the technical performance of cars contribute to an increase in noise and vibration. It is therefore not a luxury to counter these secondary parasitic effects.

ACOUSTIC COMFORT

Before examining the different techniques of noise control, it is necessary to define how noise can be judged, estimated and compared; and that is where the first difficulties begin. Two means can be used: physical measurements and subjective judgement. Experience shows that it is necessary to use both.

Based on physical measurements, two types of graphs allow the representation of most acoustic phenomena relative to car noise. These are the sound-pressure level expressed as a function of frequency (noise spectrum) and as a function of the vehicle speed or rotational speed of the engine.

Fig. 15.1 shows the noise spectrum inside a saloon car running at 120 km/h. Such a curve can be considered indispensable in defining a noise but, nevertheless, depends on knowledge of the human ear for its interpretation. For example, each dotted curve in Fig. 15.1 corresponds to a disturbance sensation balanced between high and low frequencies.

The spectrum serves as a base for the calculation of different parameters that take into account the sensitivity of the ear and characterize the noise by a single value. Currently 8–10 parameters of this kind are in use. Recent unpublished work has shown that for the

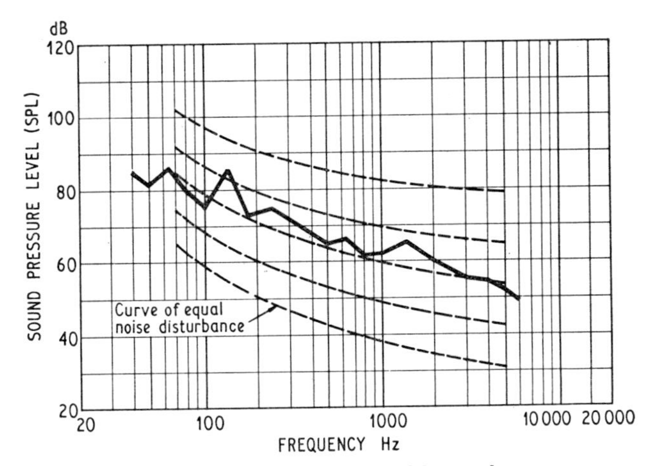

Fig. 15.1. The noise spectrum inside a saloon car at 120 km/h

appreciation of noise comfort in cars, three of them give satisfactory results.

The weighted sound level in dB (A) is the least precise of the three but very interesting because it can be read directly from portable sound level meters.

The loudness in sones is considered to be the most valuable criterion for the representation of a spectrum. It also has the great advantage that it is proportional in good approximation to the auditory sensation. This makes it possible, for example, to express the result of a noise reduction as a percentage.

The articulation index is also interesting from the practical point of view because it allows a fairly good estimation of the ease with which two people about a yard apart can converse.

Fig. 15.2 shows a typical sound-level curve as a function of the car speed. In order to assess 'comfort' the sound-level is usually expressed in dB (A). However, in numerous cases dB (B) are used because this gives a better picture of the behaviour at low frequencies, where some very disturbing maxima ('booms') are frequently encountered.

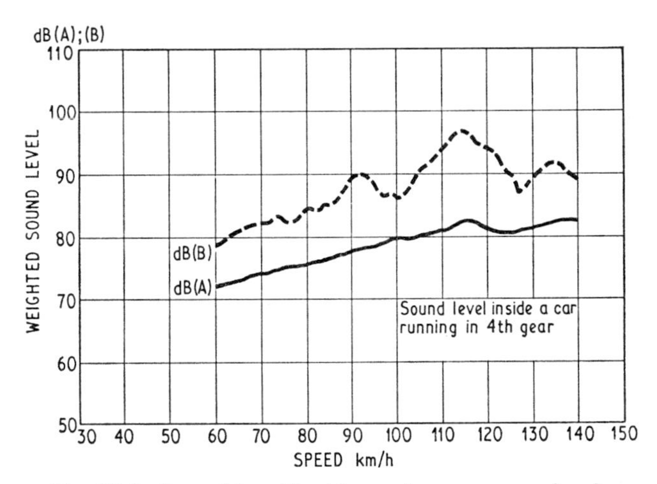

Fig. 15.2. Sound level inside a saloon car running in fourth gear

Experience has shown that in order to assess a car for noise, it is necessary to have the following data, taken at two microphone positions, one at the front, one at the rear.

The sound level in dB (A) and (B) as a function of speed in top gear.

The difference between the sound in dB (B) obtained on smooth and cobbled surfaces.

The noise spectra at normal and near-maximum speeds, as well as those corresponding to town driving.

The calculation of the loudness and of the articulation index thus permits a better comparison of the comfort in different cars.

Assessment of acoustic comfort by subjective examination is not easy. The principal difficulty is man's lack of memory for noise intensity. The observer has no reference noise level and his appreciation is greatly influenced by the physiological and psychological conditions prevailing at the moment.

That is why a comparison made by passing rapidly from one car to another after a short drive leads to the best subjective judgement. The more rapid the change-over, the easier the comparison. The use of magnetic tape recordings therefore gives excellent results.

Great progress has been made in the reproduction of noise inside cars by improvements in the electronic apparatus and in the listening conditions. Excellent results have been obtained by the judicious distribution of about 50 loudspeakers mounted on the inner surfaces of a car body.

For the car manufacturer the subjective judgement remains of the greatest importance even though it presents difficulties and cannot be replaced by physical measurements. Some cases of noise roughness, drumming or pressure sensation which can be very disagreeable are hard to define by the usual methods of measurement.

THE PRINCIPAL ACOUSTIC FAULTS

When a car manufacturer wishes, or is obliged, to improve the acoustic comfort of a model, it is of primary importance to define the faults as exactly as possible. They can be divided in two groups.

Faults appearing in well-defined running conditions, for instance a noise maximum at a certain car speed or rotational speed of the engine. The common characteristic of this group is the existence of resonance between the noise or the vibration sources and the sound pressure near the ear of the passenger. Such a fault will be referred to as resonance or boom.

Faults of a more general nature such as a high overall sound level, too much engine or rolling noise, and loud wind noise at high speeds.

Often these two kinds of faults are superimposed. Their separation is very important because the methods for eliminating such faults are often completely different and the order in which they are treated is important.

Fig. 15.3 shows this clearly. Acoustic treatment applied to a car with resonance may reduce the sound

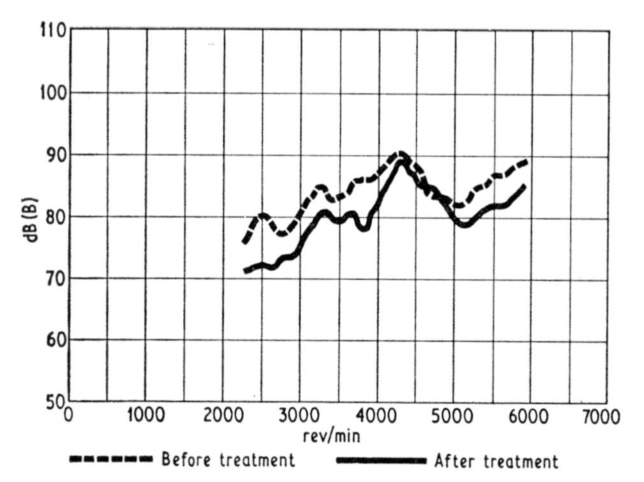

Fig. 15.3. Sound-proofing of a car with a resonance may reduce the sound levels on either side but not affect the peak itself

levels on either side of the noise peak, without changing the peak itself. The car will then be judged subjectively to be worse than before. In most cases the resonance peaks must be reduced to acceptable levels before attempts are made to reduce the general sound level.

RESONANCE PHENOMENA

These phenomena are found on nearly all car prototypes. They can be grouped according to their origins as follows.

The engine or its auxiliaries create a maximum of noise which is transmitted to the inside of the car, for example, acoustic resonance associated with the inlet filter and heard through the bulkhead.

The engine or its auxiliaries pass through a vibration maximum which is transmitted by the car body and produces a noise-peak inside the car. An example of this is the flexion resonance of an engine–gearbox unit.

Acoustic cavity resonances occur inside the car. A stationary wave can develop and produce local noise concentrations.

Parts of the structure or panels have natural frequencies which can be put into resonance.

Very often the disturbing faults have several origins and, even in cars of the same series, it is not always the same cause which is dominant.

The noise and vibration maxima of the engine and other sources are now much easier to detect because of the improvement in apparatus, and most car manufacturers are able to solve such problems in a fairly short time. The components can very often be examined separately before the prototype is mounted.

Stationary wave systems are relatively easy to detect, especially by the CO_2 test, but the possibilities of eliminating them are few and not always practicable. Much work is being done on this subject, both theoretically and practically, and it is to be expected that better methods will become available.

The situation is different for problems concerning the structure and panels; the development of the monocoque body has added to the difficulties. The rigid chassis to which the other parts are fixed has now disappeared. Today the panels play a very important part in the transmission of forces and contribute more and more to the rigidity of the body. Acoustically this has led to the following consequences.

The steel panels are subjected to high stresses and often behave in a completely unaccountable way from the point of view of their acoustical radiation. Their behaviour is very unstable and, since the manufacturing tolerances are relatively great, the result is a wide spread from one car to the next.

The vibration sources (engine, transmission, suspension) are often fixed to pseudo-frame members of low rigidity; consequently the apparent mass at the mounting points is low also. In other words, since the mechanical impedance is low, there is significant transmission of vibration into the body.

The modern car body is usually carefully examined statically and also dynamically at low frequencies; but this is rarely extended beyond the beginning of audible frequencies. The 80 to 250 Hz range, however, contains important excitation frequencies from the engine and the transmission.

Previously, resonances in this range were due to flexion or torsion of a cross or frame member but with the monocoque body such resonances often involve a complex deformation of the whole body.

Perhaps the most important consequence for the car manufacturer is that when such body resonances have been laboriously detected, their elimination necessitates costly modifications of production tools.

A further aspect which complicates the acoustic problem is that faults only become apparent when the prototype is completed. Little time is then available for important modifications and, furthermore, the behaviour of prototypes is rarely representative of the production cars.

THREE REMEDIES

Three complementary techniques should facilitate the solution of such problems in the future.

First, an apparatus designed for the study of the dynamic behaviour of vibrating structures will soon be available. Named the RTAD (Real Time Analog Deformation) instrument, it functions on the following principle: the signals from 6 or 12 transducers, placed on an element of the structure, are filtered at the frequency to be studied. The simultaneously determined instantaneous value of the amplitude of movement is shown on 6 or 12 meters side by side.

The phase condition of sampling can be changed in accordance with the signal of a reference transducer, placed at the excitation source or on the element under examination. The deformation and movement of the element can thus be visualized in real time, photographed or read in absolute values. Fig. 15.4 shows the apparatus

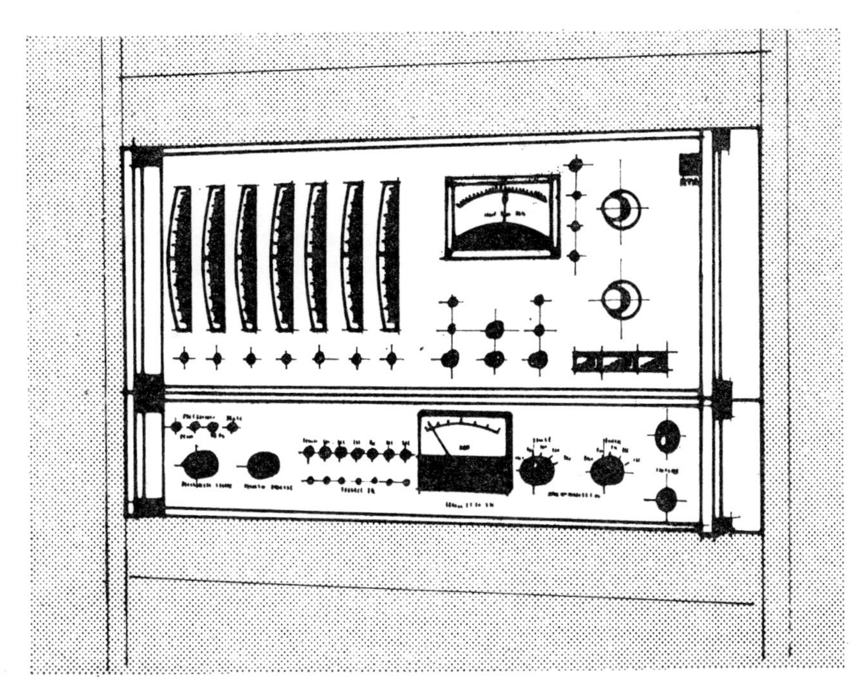

Fig. 15.4. The real time analogue deformation unit and its calibration unit: deformation can be made visible in real time

and Fig. 15.5 the deformation of a car body. This apparatus will effect an enormous saving of time and improve accuracy, compared with the commonly used method of measuring point after point.

The second technique is the study of the resonance modes of bare car body in the critical frequency zone. Knowing the excitation frequencies of the mechanical parts, one should be able to predict the modes which could be critical acoustically. This would allow an earlier intervention, before expensive production tools are set up. However, the validity of such a method needs to be proved, for the bare car body may vibrate at frequencies different

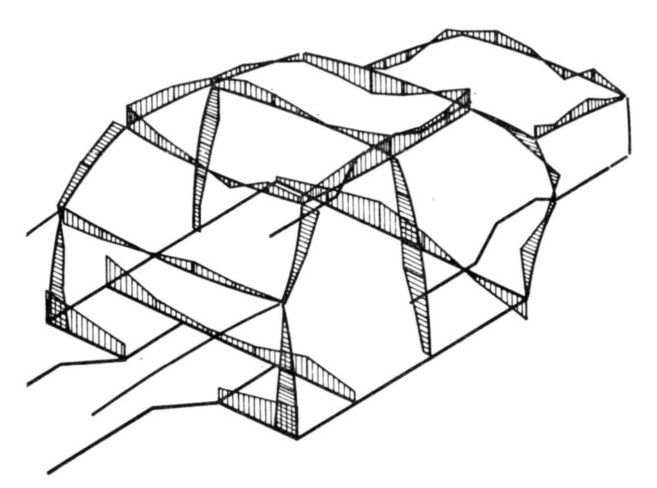

Fig. 15.5. Deformations of a car body structure for one phase condition at 140 Hz obtained by using the apparatus in Fig. 15.4

from the completed car and it is necessary to verify in each case that the vibration maximum corresponds to a noise maximum.

A third technique would demand a better knowledge of the acoustic effects of the sound radiation from the different surfaces of the interior. Experience has shown that small modifications, such as the application of supplementary weights at well defined positions, can often greatly influence the noise maxima. Modifications of the behaviour, especially the phase condition, of certain critical surfaces by mechanical or electronically assisted techniques should permit the control of the sound emission.

Before these techniques are perfected it is, however, possible to formulate certain useful practical rules which, though not infallible, reduce the chance of disturbing resonances appearing.

The transmission of forces should not create torques on the frame members.

The mounting points of the engine and of the suspension should be very rigid to make the apparent mass at these points as large as possible.

The coupling between interior and exterior mainframe members should provide continuous rigidity between the front and the back of the car.

The bulkhead should, as far as possible, be free from mechanical stresses.

THE REDUCTION OF SOUND LEVEL

In order to solve sound level problems, which are very frequent, it would seem logical to begin with the noise and vibration sources. Although much work has been, and is being, done in this field, it seems that it will be some time before the results are translated into practical remedies. A review of progress during the last 10 years shows that improvements have resulted mainly from modification of the car body rather than of the noise sources. Even the introduction of new systems of propulsion, such as the motor with revolving piston, the air-motor of Stirling and Meijer, the electrical motor and the turbine, will only affect some of the noise sources.

Leaving aside the problem of the sources, noise reduction then becomes a question of diminishing or preventing the transmission of noise to the ear of the passenger.

THE MODES OF NOISE TRANSMISSION

It is essential to make a distinction between the two modes of transmission which are found in nearly all acoustical problems.

Airborne: the sound waves emitted by the source first pass through the air on the path to the ear.

Structure-borne: the vibratory energy emitted by the source is propagated through the structure before being radiated into air by panels.

In cars low-frequency noise is transmitted by the structure whereas high-frequency noise (above about 500 Hz) is transmitted only by air.

THE TECHNIQUES OF REDUCING NOISE TRANSMISSION

The principal ways of reducing transmission are:

Vibration insulation.
Damping.
Sound insulation.
Absorption.

The first two mainly concern structure-borne noise and the last two airborne noise.

The principle and the application of each different technique need to be clearly understood in order to use it efficiently.

Vibration insulation

In most cases vibration insulation devices are located at the points where the vibrating sources are attached to the structure and consist of elastic elements, usually rubber. From the acoustic point of view, these elements should be as soft as possible and not too highly damped. However, a compromise is unavoidable because, for the principal vibration sources, engine and suspension, other considerations apply, for instance road-holding, which requires hard elements. This being a mechanical problem, it is normally solved in a satisfactory manner by the car manufacturer.

Damping

The damping treatment for the steel panels is well known and consists of spraying, glueing or baking on a layer of material with a high loss factor. Such a treatment has three main effects. It:

greatly diminishes any resonances due to the natural frequencies of the panel;
reduces the vibrations transmitted by the structure;
rapidly eliminates vibrations produced by shocks.

Laboratory studies have recently indicated that the results obtained by the usual method of measuring mechanical damping are not always valid for selecting the most effective material. It is also highly probable that increased damping of the joints connecting the different parts of the car body would reduce the structure-borne noise.

Sound insulation

Sound insulation and absorption, being purely acoustic techniques, are less familiar to car designers and are often confused. However, it is without doubt the progress in sound insulation that has contributed most to recent improvements in the acoustic comfort of cars. It is therefore worth while to consider this technique in more detail.

Sound insulation consists in reducing the transmission of airborne noise by interposing partitions so that only part of the sound energy goes to the listener. It can be obtained in two ways:

By the use of a heavy partition.
By the use of a multilayer panel called 'sandwich' partition.

These two types of panels differ considerably in their efficiency as related to frequency. Fig. 15.6 shows that since the frequencies of airborne noise are predominantly above about 500 Hz, the advantages of the 'sandwich' technique are evident.

Such a 'sandwich' is composed of alternating layers of porous material and impervious sheet. Up to a point these

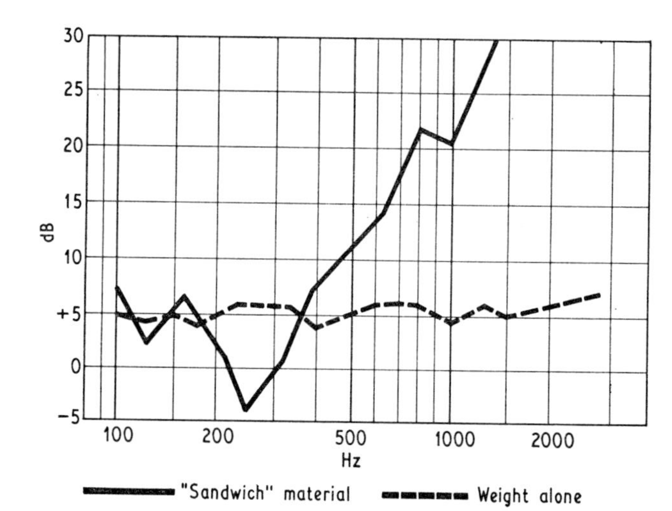

Fig. 15.6. Sound insulation by material relying on its weight alone, compared with material of similar weight but multi-layer sandwich construction

are analogous to springs and masses, respectively, in a mechanical system. Even the simplest sandwich (a single porous layer covered by a sheet) possesses a natural frequency which appears on the attenuation curve as a minimum or 'hole'. Fig. 15.7 shows the considerable variation in the insulation characteristics of different types of 'sandwiches' with identical thickness and weight.

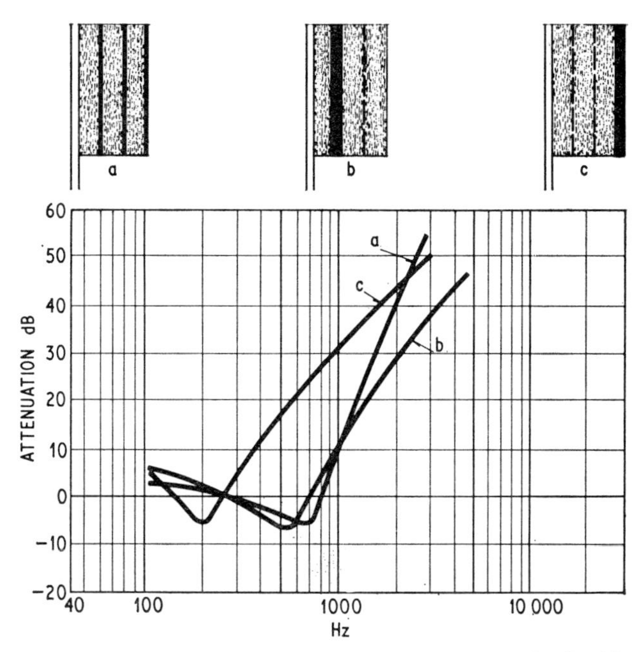

Fig. 15.7. Different sound insulation spectra obtainable from different types of sandwiches of identical thicknesses and weights

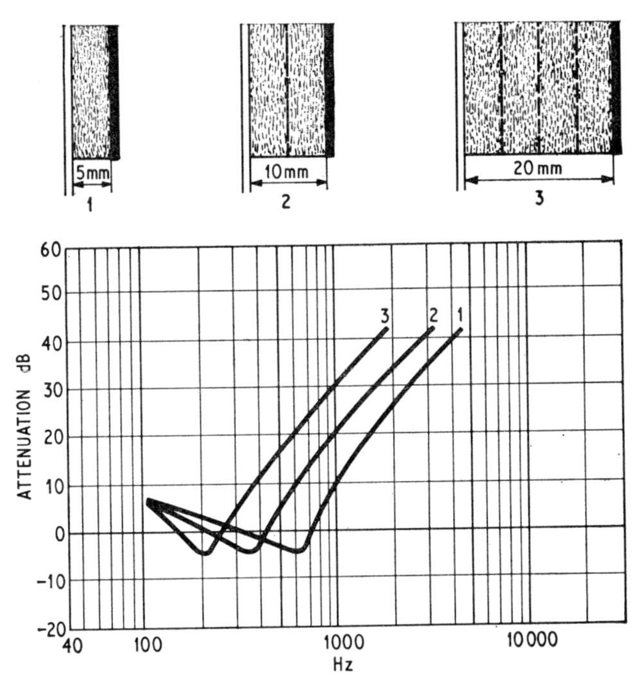

Fig. 15.8. The effect on a sound insulation spectra of increasing the porous part of the sandwich while keeping the weight of the impervious material constant

For simple 'sandwiches' which are most commonly used Fig. 15.8 shows the effect of the thickness of porous material and Fig. 15.9 shows that of the weight of the impervious layer. It is evident that the acoustic and physical properties of both materials also influence the efficiency of the 'sandwich'. Thus it is possible considerably to

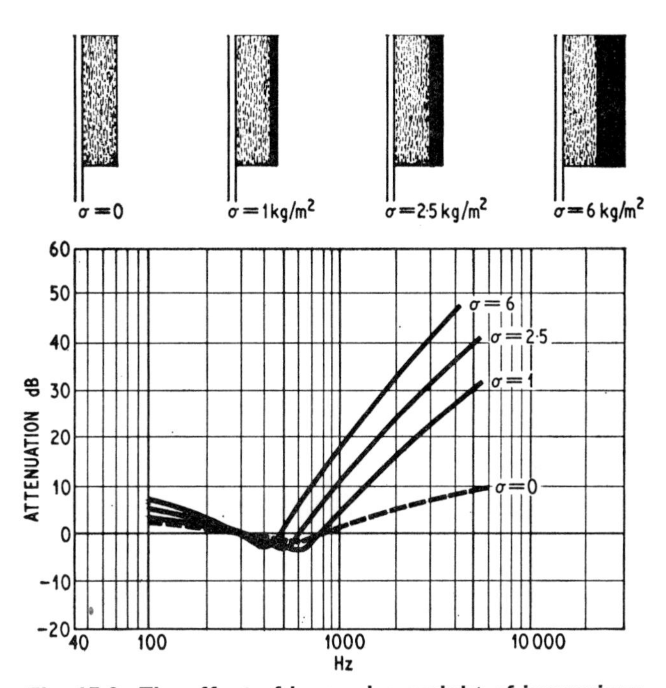

Fig. 15.9. The effect of increasing weight of impervious material at constant thickness of porous material

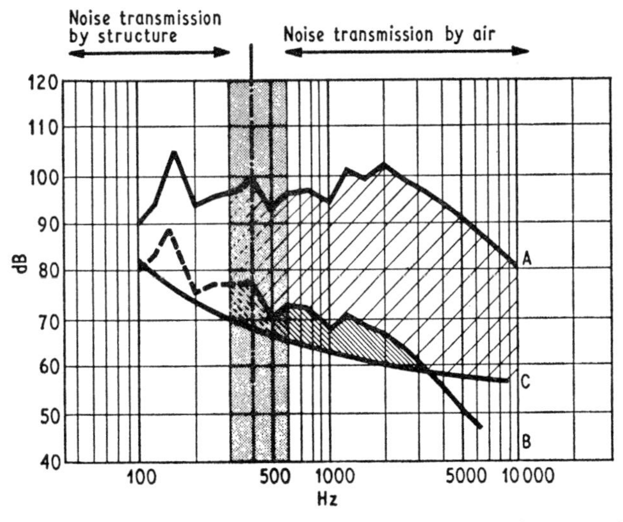

Fig. 15.10. To find optimum frequency range for sound insulation from engine draw noise spectrum in engine compartment (A) and deduct sound insulation effect of bulkhead (calculated from mass law) to obtain (B). If a curve of equal annoyance to the human ear (C) is plotted at any desired level, the range to be insulated will be enclosed between (B) and (C)

modify the insulation characteristics as a function of the frequency.

The optimum curves for materials to be used in cars have to be found. The insulation for the bulkhead, considered as the partition between the engine compartment and the passenger space, is a typical example, illustrated in Fig. 15.10. Curve A is the noise spectrum measured in an engine compartment and from this has been subtracted the attenuation obtained by the bare bulkhead (based on the mass law). The resulting noise spectrum B corresponds approximately to the noise inside the car coming from the engine. In the same diagram, curve C corresponds to the equal annoyance suffered by the human ear. The comparison of curves B and C shows that supplementary insulation is mainly needed in the middle frequencies.

From the properties of 'sandwiches' previously mentioned it becomes evident that material having the 'hole' of its frequency characteristic as low as possible in the frequency range will give the best results. It is advisable, however, that the hole be situated above the second order of the rotational frequency of a 4-stroke, 4-cylinder engine.

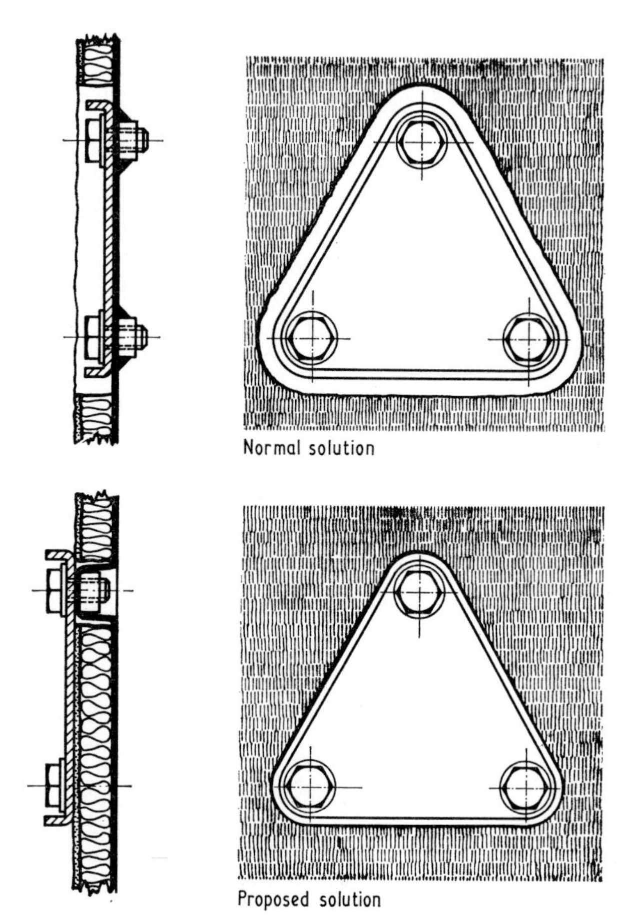

Fig. 15.11. The adverse effect of holes cut into acoustic material for fixing purposes is underestimated: here is one way of reducing the size of the hole

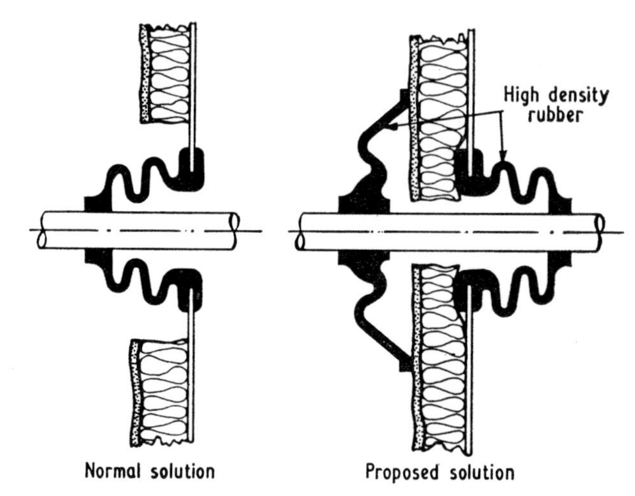

Fig. 15.12. Another method of reducing the size of hole that has to be cut into sound insulation

In practice a simple 'sandwich', consisting of a highly efficient porous material, 18 mm thick, covered by an impervious and flexible sheet weighing 4 kg/m^2, is an excellent compromise. However, the results obtained from materials shown to be good in laboratory tests are often disappointing. The cause of this discrepancy lies in the influence of numerous 'acoustic holes' which exist on all cars. These are surfaces that have an insulation distinctly lower than their surrounding surfaces. Some examples of 'acoustical holes' are:

An actual hole in the steel sheet of the bulkhead.
The rubber sleeve of the brake or gear lever.
The windows, when their insulation is lower than that of the rest of the body.

In the bulkhead, two kinds of acoustic holes are found: the physical holes necessary for the passage of the command elements and the joints between the different pieces of acoustic insulation. Care is necessary at the design stage to limit the holes to a minimum and ensure that all the passages through the steel sheet are air-tight. A further precaution is to reduce to the minimum the size of the holes cut in the acoustic materials. The influence of such holes is systematically underestimated. Two examples showing means of accomplishing this are given in Figs 15.11 and 15.12.

The second type, the joints between the pieces, contribute to the noise transmission in an unexpected manner. It has been found from theory and confirmed by laboratory tests that the influence of a slit is much greater than that of a round hole of equal area. Whereas round holes influence mainly the high frequencies, slits have their maximum effect in the middle of the spectrum. The problem can be solved by the use of large, preformed, single pieces which have recently become available commercially. Their application is simple because they can be fitted at an early stage of assembly.

Absorption

This last technique of noise reduction consists of diminishing the reflection of the sound waves by covering certain inside surfaces of the car with an appropriate material. The car roof is ideal for this kind of treatment since it is a large surface near to the ears of the passengers.

The technique of absorption is rarely used correctly. In fact, even a good absorbing material, 15 to 20 mm thick, is efficient only at high frequencies. In order to extend its efficiency into the middle frequencies, it would be necessary to increase the thickness considerably. However, if the 15 to 20 mm thick material is covered by a membrane with perforations covering 4 to 8 per cent of the surface, it is possible to obtain a peak absorption at about 1000 Hz. The loss of absorption in the high frequencies due to the decrease in absorbent surface is normally compensated by the absorbing effect of carpets and seat coverings.

An interesting aspect of this treatment is that it is independent of the noise sources and thus allows a reduction of noise from sources which are otherwise very difficult to treat, for example noise from tyres or wind on the car body.

THE APPLICATION OF THE ACOUSTICAL MATERIALS

It is important to use materials that are well adapted to car problems. It is even more important to know at which places the different materials should be applied. For a treatment to be correct, or in other words to yield the best compromise between efficiency and cost, it must be homogeneous. Thus it is useless to treat a certain area extensively while others are neglected.

As a means of obtaining such a homogeneous treatment, the subtractive method gives excellent results. Preferably carried out on an acoustical test bench, this technique consists of applying an intensive treatment, based on auscultation and experience. All ‘acoustic holes’ must be eliminated and all must end up contributing equally to remaining noise. The relative efficiency of each element of the treatment can then be accurately determined by removing one element at a time and measuring the resulting noise increase. Depending on the proposed target and the economics, a rational treatment can then be devised.

It seems probable that the next step in the improvement of the acoustic comfort will be accomplished by a more efficient use of the different sound proofing materials.

Paper 16

MODERN METHODS OF TRANSLATING A STYLING MODEL TO A BODY DRAFT AND TOOLING

D. W. Davy*

This paper is concerned only with skin panels. A system centred on a digital computer is suggested though it has not yet made of having implemented the total system. Parts of the system have been completely implemented. The computer has proved of use in other areas to relieve the designer of routine and time-consuming tasks, such as plotting wheel movement envelopes and as an aid to the design of small press tools for internal panels, but this subject is not covered here.

INTRODUCTION

PAST PRACTICES of translating the exterior skin shape from a stylist's model to a body draft have proved to be time-consuming and tedious. The traditional method was cutting and fitting templates to the clay and using these as a basis for the body description. This involved a skilled labour force and was an inflexible system, prone to error. Many problems were experienced in the drawing office in aligning the templates to a common datum, due to their relative inaccuracy.

More recent methods have involved using a 'bridge' for manual measurement of co-ordinates on the clay as a basis for laying out the initial draft. Though an improvement on the template method, this has a number of limitations: the co-ordinates are read from scales which was time-consuming and the vertical datum varies as the bridge moves along the body.

With both these methods the designers concerned with feasibility studies and scheming the interior panels and structural members had to wait until the initial lines were laid out and a print taken, before they could effectively begin their work.

COMPUTER METHOD

Mathematical model

Modern methods of translating the shape of a clay model into the body draft centre on the use of digital computers.

The MS. of this paper was received at the Institution on 1st July 1970 and accepted for publication on 3rd July 1970.

* *Section Head, Pressed Steel Fisher, Research and Development, Cowley, Oxford.*

The crux of the process is entering, and holding, an accurate description of the body shape in the computer. When a method of doing this has been devised, programmes may be written to present information to designers working in different areas in a detailed and meaningful way so that they may perform their functions efficiently.

Various approaches have been used to obtain a description of the exterior shape within the computer (**1**)—(**4**)†. These may be considered as falling into two main categories.

(1) Providing the stylist with a computer system which he will use to aid the styling of the vehicle, thus obtaining a mathematical model of the exterior surface at the conception stage.

(2) Allowing the stylist to complete his work conventionally, producing a clay model and using this as a basis for obtaining a mathematical model of the surface.

The problems associated with the first approach are mainly those of size and medium:

(*a*) Apparent changes in proportions when a shape styled to a small scale is directly scaled up.

(*b*) Working with a two-dimensional medium, such as a cathode-ray-tube or drafting machine, does not give a representation which allows good three-dimensional visualization (the screen is also too small). This has been found to be true, even when rotation and perspective facilities have been provided.

(*c*) Working with a drafting machine and numerically controlled machine tool (**4**) presents problems at the

† *References are given in Appendix 16.1.*

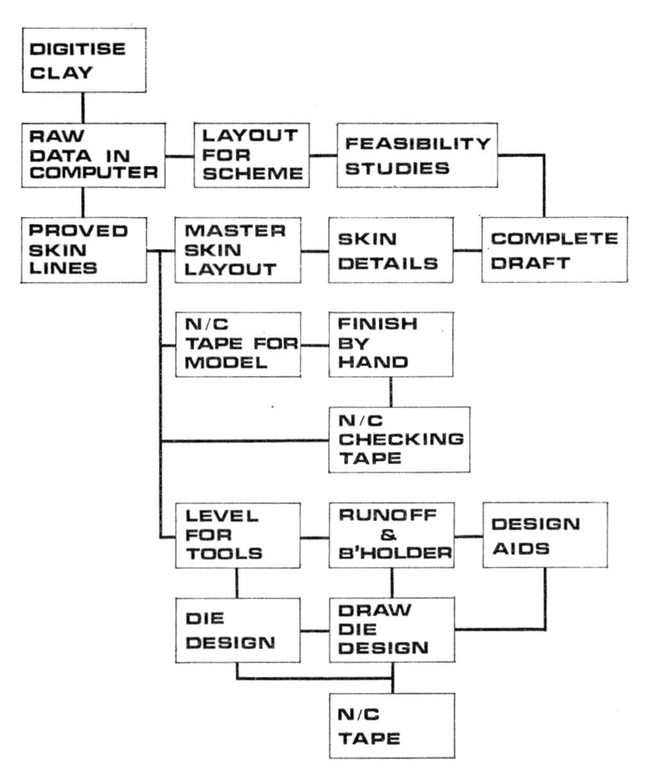

Fig. 16.1. An integrated tape preparation system which begins with a clay model produced by the stylist

styling stage in splitting cars down into areas for machining. However, this general approach is the one most likely to succeed at the styling stage.

(*d*) If a model of the final surface is to be set up within the computer, some surface proving will have to be done at this stage. This is normally the first stage of body engineering.

Clay model

An approach to an integrated system, working from a clay model, is suggested in Fig. 16.1. The problems of integration into the normal method are not as acute with this approach as with the first. It does, however, present the difficulty of having to retain the stylist's ideas (as expressed in the clay model) whilst fairing the surfaces in the computer. The computer is used as common data base and link between components of the system, ensuring accuracy and consistency of information.

The information starts with the full size clay model produced by the stylist. Co-ordinates describing sections and character lines are taken from the model, with the aid of a large automatic measuring machine, and recorded in punched paper tape (Fig. 16.2). This unrefined information is fed to the computer and structured in a data base so that any area may be easily accessed.

The information held in the preliminary data base may be viewed and checked, using a graphic display on-line to

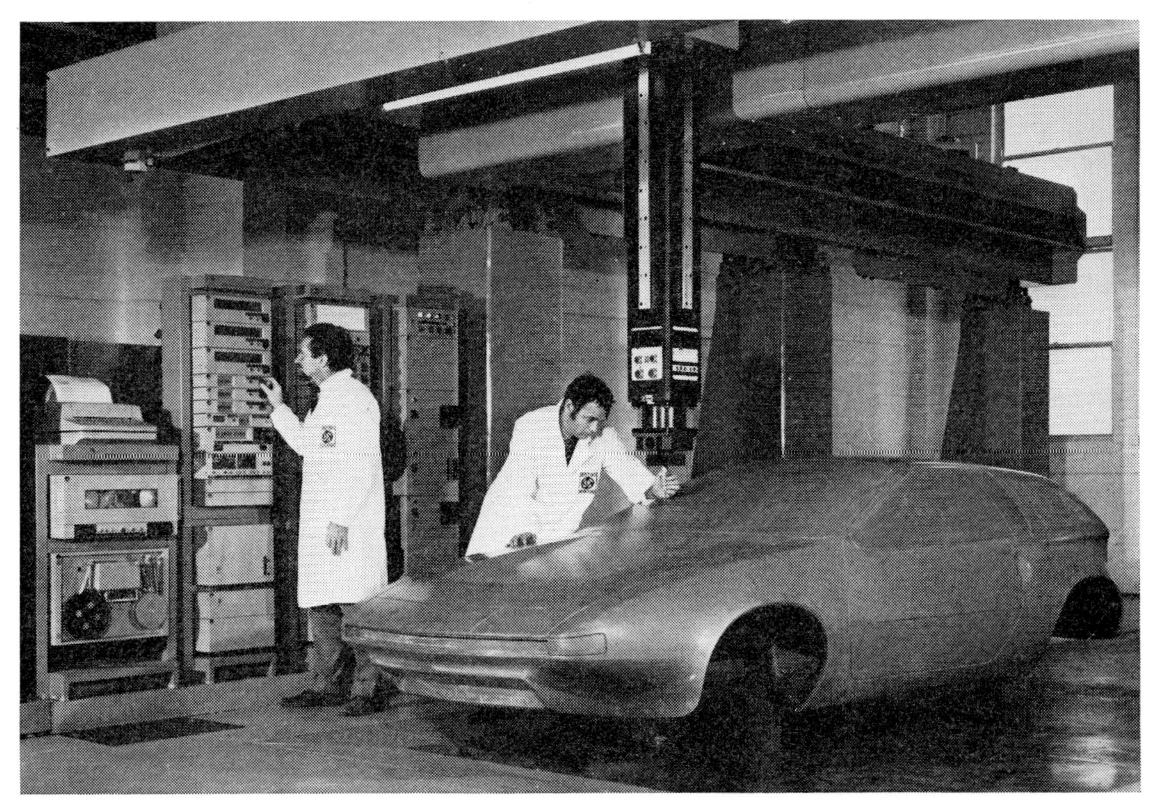

Fig. 16.2. Co-ordinates are taken from the model by a hand-operated stylus and automatically punched on paper tape by the equipment

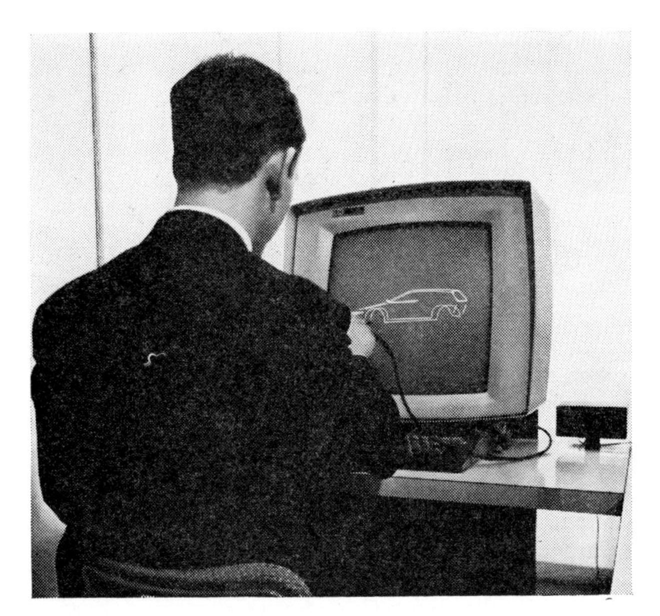

Fig. 16.3. Co-ordinate information held in the computer may be viewed in graphic form on this CRT display

the computer (Fig. 16.3). Any combination of areas may be viewed on the screen, they are selected by pointing with a light-pen to a list of area names displayed on the screen. Obvious errors may be corrected by indicating, again with the light-pen, the co-ordinate to be modified, and then typing in the new co-ordinate on the keyboard. Various facilities are provided, such as enlarging, selecting views, etc., to ease the task of assessment and modification.

When the raw data have been monitored and obvious errors corrected, initial layouts may be requested, via the screen, of any combination of areas. These layouts are then produced by the automatic drafting machine (Fig. 16.4) under control of the computer and may be used in the scheming of exterior panel splits and the planning of interior panels and load-bearing members. Separation of views, resting sections and other similar functions are performed by the programme (Fig. 16.5).

Limitations

The next operation is to refine the raw data and produce a description of the proven exterior surface. The mathematical and computational techniques required to 'fair' the surfaces used in modern automobile styling are complex. It is unlikely that the complete process of proving the skin lines can ever be performed completely automatically by a computer programme, because of the imponderables involved.

There are many functions in proving a surface, such as producing light lines and retaining certain characteristics, which must be based on subjective judgements.

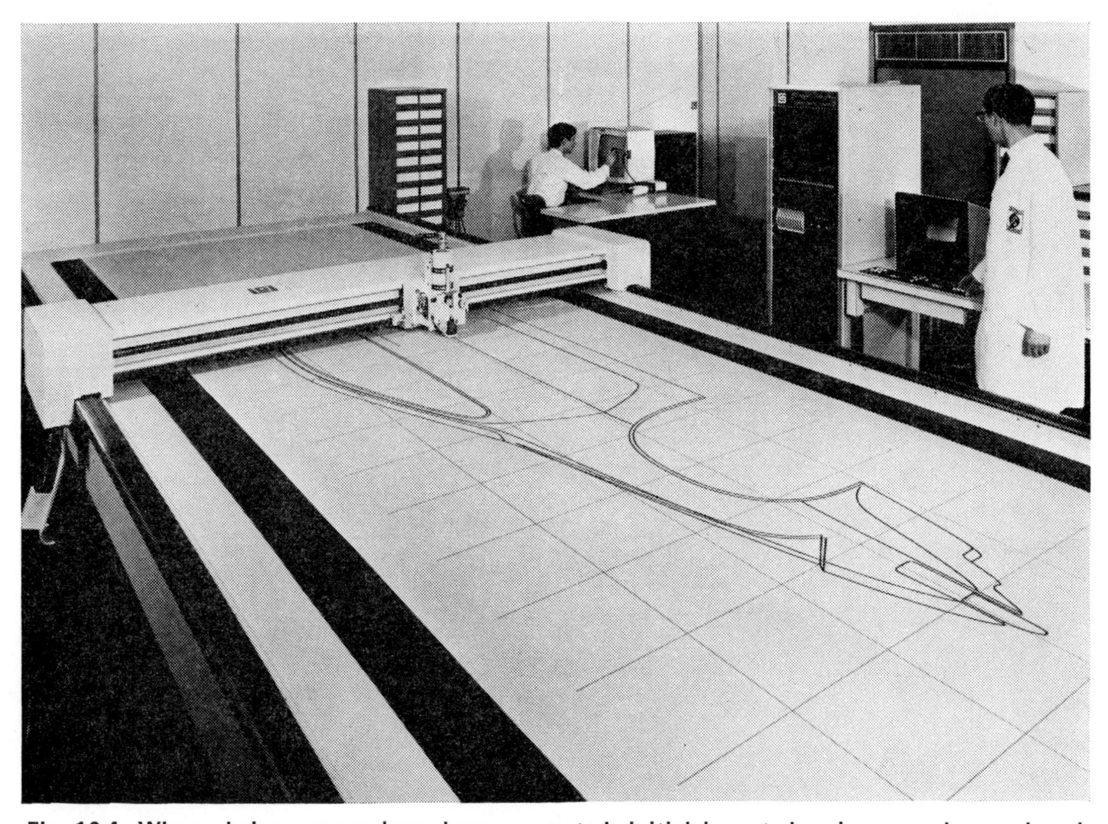

Fig. 16.4. When obvious errors have been corrected, initial layout drawings can be produced automatically on this draughting machine

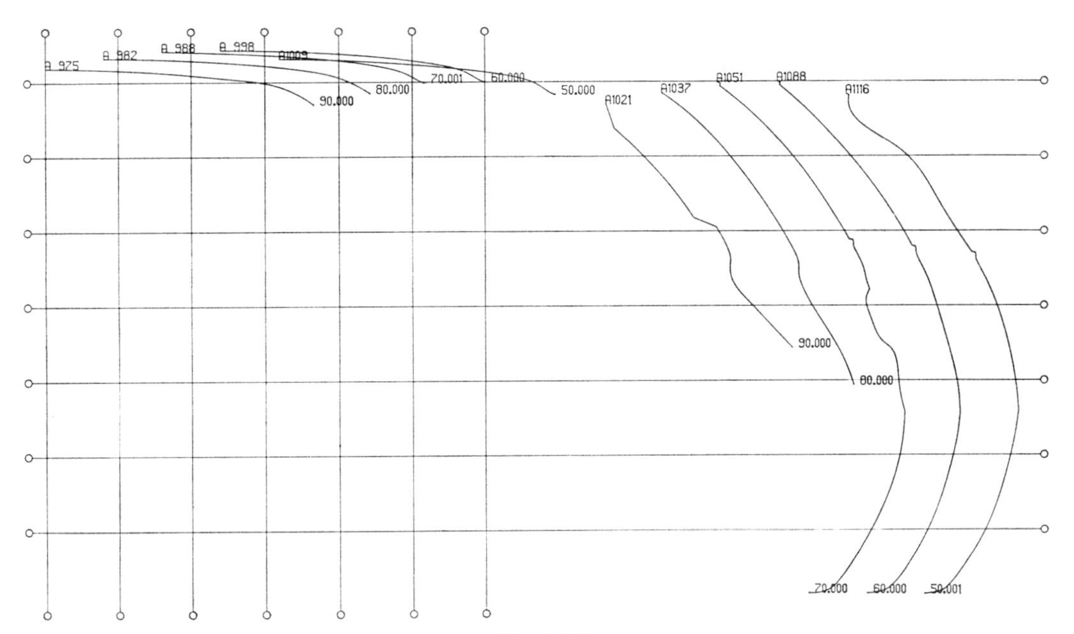

Fig. 16.5

A viable system would be one in which the computer programmes perform these functions and the engineer is allowed to assess the results and modify the surface. Such a system would probably involve the use of the graphic display and drafting machine, used interactively in the same programme.

Although the graphic display is versatile its small size and relative inaccuracy make it unsuitable for viewing faired body lines. The results of the fairing programmes would be presented for analysis and decision by using the automatic drafting machine, its large size and accuracy making it suitable for this purpose.

The display provides a convenient means of requesting information, guiding the programme by selecting alternative courses of action and entering any required change into the computer.

Thus in the complete system the graphic display and drafting machine are both under the control of the programme: one being used interactively to control the operations performed on the surfaces and input/output, whilst the other is used to present, when requested, the current shape of the surface for review.

To complete the mathematical model, the lines describing the splits between panels must be defined and incorporated in the data structure.

Master layout and details

When the computer contains an accurate model of the final exterior shape of the automobile, the major hurdle of implementing the total system has been passed. Using this model as a master data base, an integrated system may be developed.

Referring back to Fig. 16.1 it can be seen that application programmes in both body and tool engineering functions may be written to work from the data base.

A master layout of the skin lines may be produced on film or a 'metal' draft, using the automatic drafting machine. As with the scheming draft, the presentation is handled by the programme, selected information being provided on separate drafts, if required. These may then be used in the conventional manner, laying in the flanges, structural members and internal panels. Details of each individual skin panel may be produced for use as reference in the Engineering Department and in areas such as Process Planning.

The skin details may also be used by the model makers to facilitate 'blocking up' for the production of models. As the system has refined the data taken from the clay, a full-size model is required for final viewing and formers for prototype manufacture. The data in the computer may be used to produce an N.C. tape for machining the surface on the blocked-up models.

This machining operation will inevitably leave some excess material, however small, in the form of cusps which have to be removed by hand. The same data used to produce the machining tape could be used for a tape which drives the automatic measuring machine, for checking the final surface.

The facility of producing machining information for models would also be required if the system were used at the styling stage (though the N.C. machine tool would probably have to be on-line to the computer in this case).

Tool manufacture

The next stage would be the use of descriptions held in the data base to aid the design and manufacture of press tools. The requirements for surface manipulation to aid tool design differ from those for skin proving. The skin panels are now precisely described and the surfaces need to be modified and re-oriented to meet manufacturing requirements. A panel may have to be oriented differently in space for each operation—such as draw, trim, flange, etc. —to proceed from the flat sheet to a finished part.

For the various operations, the panel will have to be progressively modified to provide conditions suitable for each stage of manufacture, e.g. flanges opened for the draw and trim tools. The orientation of the panel in space will also change to facilitate each operation.

The initial task to be performed in this phase of design, therefore, is to re-orient, modify the boundaries and other areas of the panel, and construct addendum areas (such as blank-holder surfaces) with regard for the requirements of the various pressing processes.

Each panel is being considered separately so that the small size of the screen becomes less of a restriction, especially if the facility for enlarging local areas is provided. Because of the speed of modern computers, surfaces may be presented in a way which enables the tool designer to visualize the shape more easily and thus better judge the problems of manufacture. An example of this is a surface presented by a series of parallel, closely spaced sections through the piece (Fig. 16.6).

Full or partial sections may be requested to show any area so that the tool designer may easily examine the panel in detail, undercut conditions (affecting the orientation of the panel) may be found by using these facilities, or possibly by automatic methods.

The panel may be rotated about any axis on the screen and, in conjunction with the examination procedures, this helps to decide its orientation for different manufacturing operations. Facilities may be provided for local modifications of shapes, the designer using the screen to indicate modifications required and the computer retaining an accurate description of the panel.

An extension of these facilities, providing for surfaces to be developed outside the boundaries of the panel, would allow run-off surfaces and blank-holders to be developed. This would mean that the tool designer describes a number of significant sections, defining these areas in terms of certain points, and the programme would fit a surface containing these sections.

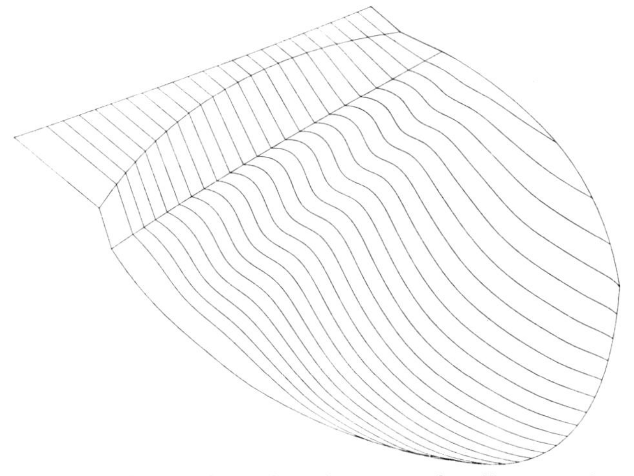

Fig. 16.6. To enable tool engineers to visualize manufacturing problems, shapes may be presented as parallel, closely spaced sections by the computer controlling the draughting machine

The examination procedures could allow the designer to look at the surface thus generated and decide if modifications were necessary. If so, he would modify the shape of the sections previously defined and repeat the surface generation procedure, the designer making decisions based on judgement and the computer performing the calculations required to produce a surface from his sparse data.

If a complex panel is encountered and, even with the aids described, conditions are difficult to visualize, a three-dimensional model may be required. This can be achieved by producing a machining tape from the mathematical model which is used to produce a three-dimensional design aid in a material such as polystyrene. When an acceptable surface has been defined, the model held within the computer provides the starting point for tool design.

Some parts of a press tool are controlled by the size and shape of the part being produced, others are standard or may be designed using easily defined rules. Apart from the piece shape, therefore, all standard items to be used in the tool must be stored within the computer.

A set of programmes containing algorithms for design and logical selection of components must also be written. Using these as building blocks, programmes may be written to aid the design of the press tools. Experience has shown that usually the best approach is to write separate programmes for different types of tools. Although there are some functions which are common to a number of tools, most of the design algorithms are different. Thus a general system for the design of any tool would probably be unwieldy and inefficient.

The general structure of a computer-aided tool-design system is shown in Fig. 16.7. The system analyst identifies those parts of the total cycle which are routine and those which are subjective judgement—based on experience or training—and obviously cannot be performed by a programme. The latter usually require the results of the routine functions as a basis for decision making.

The figure illustrates the use of the graphic display as a link between the routine programmes, this integrating the otherwise separate programmes into a single system.

All standards: stock materials, press equipment and standard items are stored within the computer and are available on request for display on the screen, or may be automatically selected by a programme. Algorithms may be written to perform the routine functions which have been identified for the particular family of tools covered by the system. Graphical console procedures allow the engineer to guide the programme through its various stages, choosing from available options when appropriate and providing a means of information input following subjective decisions.

A variety of techniques are used in these systems: facilities for design drafting, selection and placement of standard items, selection of alternative functions at various

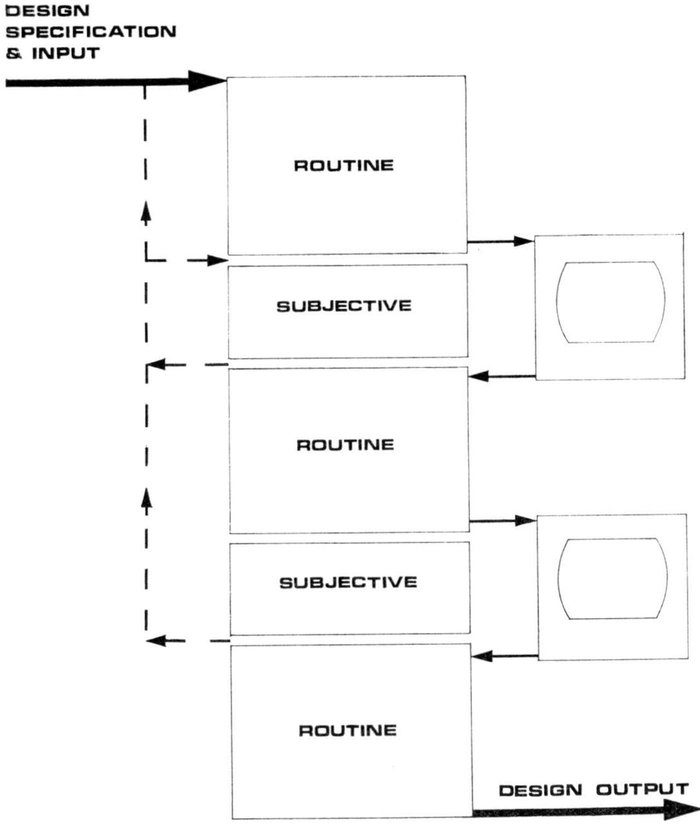

Fig. 16.7. The structure of an interactive, computer-assisted design programme for tools: subjective parts must be done by humans but are based on routine computer information

stages and examination of panel conditions. The ability to store the current stage of a design and retrieve it later for continuation is also provided. The output from these systems is usually a working drawing, produced under the control of the computer by the automatic drafting machine (Fig. 16.8). The programmes write instructions on magnetic tape in a drafting language which is later interpreted by another programme which controls the movement of the drafting machine.

Where possible, the geometric definitions describing the objects which constitute the tool assembly are stored in the computer for later use by numerical control programmes. This leaves the part-programmer to add the machining technology, and relieves him of the task of taking geometric information from a drawing and converting it to a form suitable for input to a computer.

Figs 16.9 to 16.12 are illustrations of some of the facilities previously described which are provided in current working systems. The examples are not all taken from systems applicable to skin panels but do illustrate some of the techniques required.

Fig. 16.9 shows a view of the screen taken from a programme in a design/drafting phase. The current stage of design is shown in the working area and a 'menu' at the bottom of the screen indicates available options for defining a line relative to a point. On the right an area has been enlarged to show greater detail. (The example is taken from a programme to aid the design of progression tools.)

Fig. 16.10 shows a 'menu' of available courses of action at a certain stage in the design of a single-action drawing tool. The designer may indicate with the light-pen which

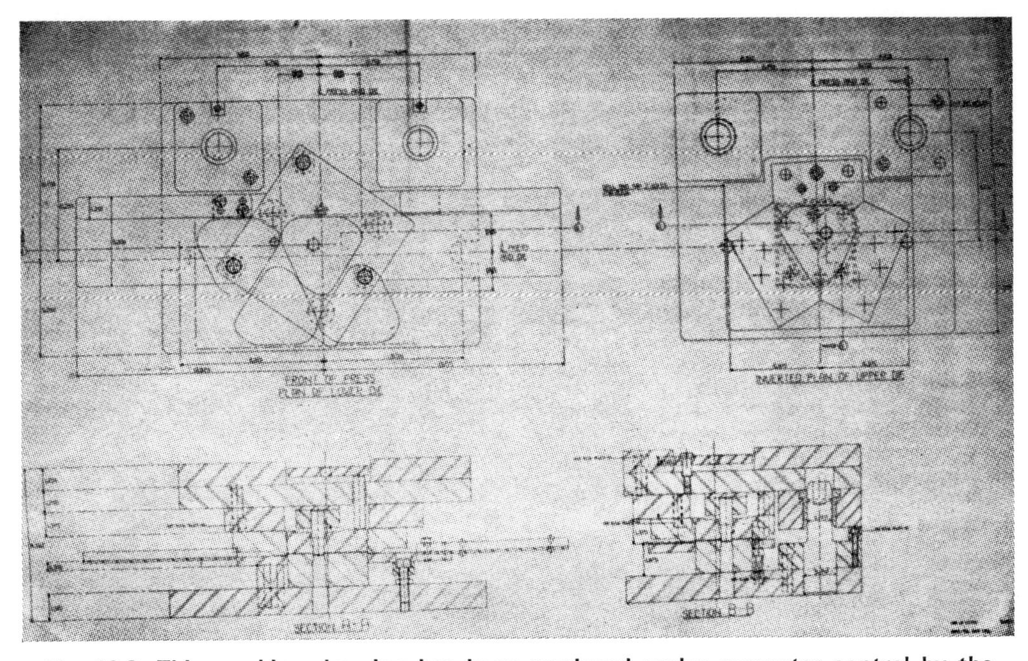

Fig. 16.8. This working drawing has been produced under computer control by the draughting machine

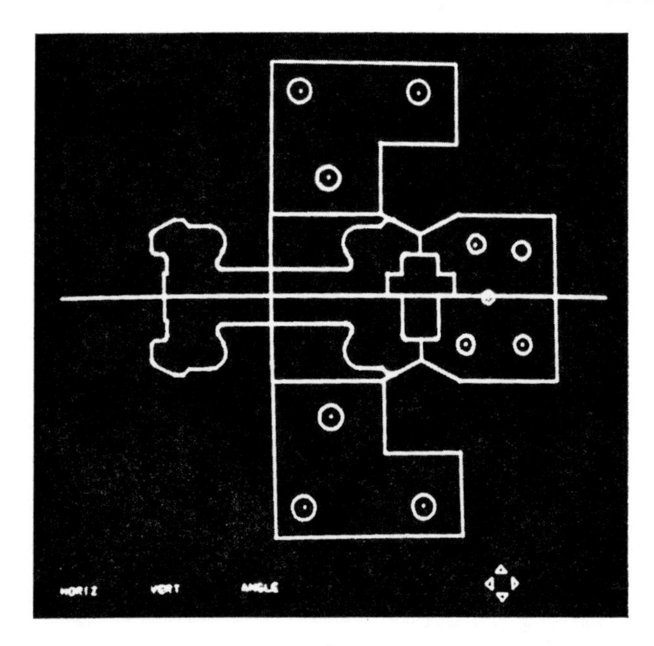

a

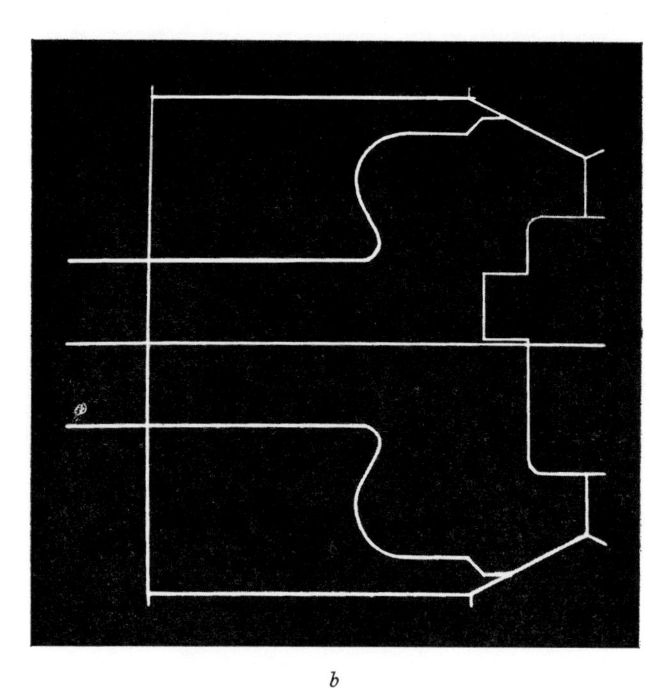

b

Fig. 16.9. (*a*) How the current state of a tool design can be thrown on the screen with options printed at the bottom and (*b*) chosen parts enlarged

alternative he wishes to pursue and a particular programme is then loaded into the computer to allow him to continue.

Fig. 16.11 shows the placing of standard components, using the light-pen on a small blanking tool. Fig. 16.12 is a section taken through a single-action drawing tool with the piece condition shown.

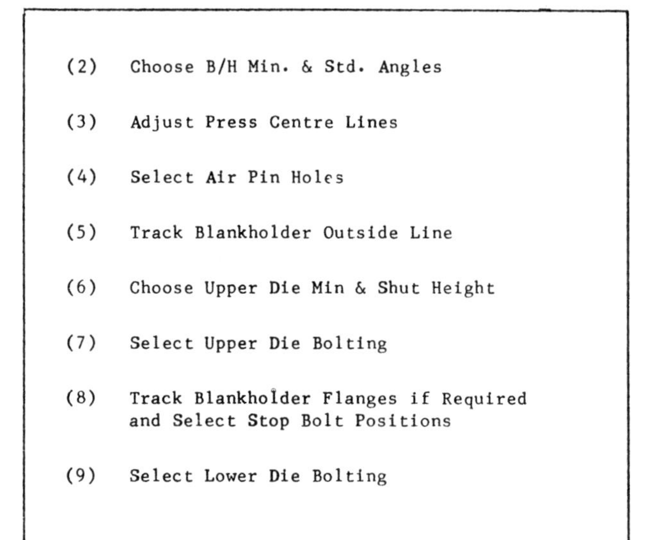

(2) Choose B/H Min. & Std. Angles

(3) Adjust Press Centre Lines

(4) Select Air Pin Holes

(5) Track Blankholder Outside Line

(6) Choose Upper Die Min & Shut Height

(7) Select Upper Die Bolting

(8) Track Blankholder Flanges if Required and Select Stop Bolt Positions

(9) Select Lower Die Bolting

(1) Cancel

Fig. 16.10. A 'menu' of options at a certain stage in the design of a tool is thrown on the screen and the designer need only point at his choice with a light-pen

Fig. 16.11. How standard components can be placed with the aid of a light-pen on the display of a small blanking tool

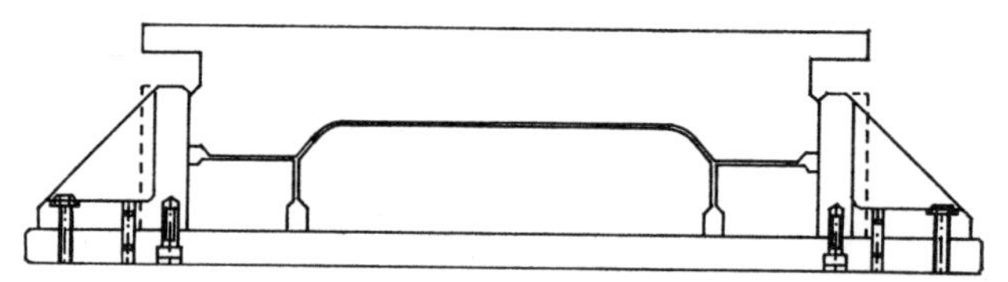

Fig. 16.12. A section through a single-action drawing tool, with the workpiece shown, is presented by the computer

CONCLUSIONS

The approach suggested in this paper has evolved from experience over a number of years of computer-aided design applied to automobile bodies. The approach of beginning with a clay model is used not because it is ideal, but because the most important factor in designing a system is not necessarily the conceptual elegance but the likelihood of its successful integration into the production system. If the system can be designed for integration and can also be extended to a more sophisticated approach, so much the better.

The advantages of using a computer are:

(1) Master information in a central and easily interrogated place.

(2) Consistency and accuracy of information.

(3) Mathematical techniques which allow the user to request any information describing the surface.

(4) Only information relevant to a function is given.

(5) A reduction in time—from release of the styling model to the production of tools—may be achieved.

(6) An integrated design/tool manufacturing system for skin panels.

(7) A reduction in design and manufacturing costs is possible.

Some parts of the system described have been implemented at Pressed Steel Fisher, working from data provided by clay and final models and design aids. Development is proceeding to provide the other components of the system.

APPENDIX 16.1

REFERENCES

(1) Coons, S. Surfaces for computer aided design of space forms (M.I.T., Project MAC, June 1967).

(2) Johnson, W. L. *et al.* Analytic surfaces for computer aided design (S.A.E. paper no. 660152, January 1966).

(3) General Motors D.A.C. I system.

(4) Bezier, P. E. How Renault uses numerical control for car body design and tooling (S.A.E. paper no. 680010, January 1968).

Discussion

Dr A. H. Chilver—May I, as Head of the Cranfield Institute of Technology and in Professor Ellis's absence in North America, extend to you a very sincere welcome from my colleagues. As you know, this symposium, and the earlier ones, have been organized jointly by the Automobile Division of the Institution of Mechanical Engineers and the Cranfield School of Automobile Engineering. Our Advanced School has always welcomed an opportunity of helping in the organization of such symposia and, as our automobile studies in Cranfield develop and grow stronger, I am sure that the staff will want to continue to play a very full part in these meetings.

Since your last meeting, Cranfield itself has been given a Royal Charter to develop essentially as a postgraduate university in technology, and I thought it would be pertinent to say a very few words about the effect that this may have on the Institute.

The aim will be to serve industry through postgraduate teaching and research. Postgraduate teaching will concentrate on the education of those with industrial experience.

Cranfield will broaden its range of studies to serve industry more generally. It already offers studies ranging from applied science to engineering and management and it is on this base that we intend to build.

For example, many areas of the work of the Institute can help the automobile industry, in addition to our present automobile engineering studies.

Your meeting is an international one and delegates have come from North America as well as from European countries. We welcome you, not only to Cranfield but to this country generally, and we hope very much that you enjoy your visit.

Before I ask Mr Barber to give his opening address, may I wish you all a very successful meeting and a very pleasant stay in Cranfield.

Mr Palmer—I am not going to address you, but to say that this closes the symposium. On behalf of the Organizing Committee, I would say that we hope and we think that it has been worthwhile and successful. It is the first time that a symposium on this subject has been held jointly by the Institution of Mechanical Engineers and the Advanced School of Automobile Engineering here at Cranfield. The numbers in which you have come and the contributions you have made to the discussion convinces us that it has been very well worthwhile, and we all feel that it is something that should be repeated in the not too distant future—perhaps in three or four years' time.

I would like to thank you all for coming. I would like to thank the authors for all the trouble they have gone to in preparing papers. I thank those who have taken part in the discussion, and particularly our overseas visitors who have come such a long way, Mr Sherman in particular, who gave us that interesting little discourse. I would also like to thank those who so kindly came forward from the industry and provided the exhibits downstairs: and—most important—Mr Glynn, Mr Lonsdale and other staff of the Institution of Mechanical Engineers as well as the Cranfield staff for all they have done.

Mr N. H. Watts—I would like to ask Mr Ennos two questions: first, where and how is the deceleration measured on the vehicles? Secondly, I am puzzled by his remarks on the filtering, because if you pass a time-varying signal through a 20 Hz filter, you need at least 50 ms, so that, surely, the lower the frequency band, the longer it will take to get one cycle.

Mr A. C. Hill—Mr Ennos referred to the windscreen glass coming out and commented that it was unbonded. The glass panel is normally ejected because the surrounding structure is deformed under the impact. Is he suggesting that, to withstand the barrier crash test, bonded windscreen panels are essential?

Dr G. M. Mackay—Has Mr Ennos tried simulating an intersection collision in which both cars are moving, by putting the bullet car on casters so that it does not have to move in the direction in which it is pointed? This procedure might produce a more realistic simulation of an actual collision.

Mr R. D. Lister—I am worried by Mr Ennos's earlier statement when he seemed to imply that all intrusion during side impact was bad. Intrusion, up to a certain point, is a good thing. In accidents with severe side impacts where severe injuries occur, roughly one in five people are injured by crushing between the intrusion and some interior part. That still leaves 80 per cent who are hurt by being flung to the inside. Therefore, there are two requirements. You have to provide as much padding as you can inside—and you have got to restrict extreme intrusion. I do not think the aim is to stop intrusion altogether as this would increase the amount of interior padding that is needed. Some compromise between these two requirements is necessary.

As regards his earlier remark about vehicle/barrier impact results agreeing with the vehicle/vehicle impact for identical but not for different vehicles, should one not consider the change of velocity, instead of the nominal velocity at impact?

Mr N. H. Watts—Mr Curtis talked about putting the mass/spring model of the vehicle under acceleration and avoiding springback. What about damping?

Mr C. C. Norville—I wonder if Mr Barley could tell us how important the transmission line is on front engine–rear wheel drive vehicles during frontal impact, since his model almost suffered total collapse in the 30 mile/h model collision?

Would he not go as far as to say that front-engine vehicles with rear drive are safer than rear-engine vehicles?

Mr G. H. Tidbury—If a theoretical model is used with a linear spring stiffness to simulate the crash properties of the material it is necessary to put a stop in the model to prevent rebound. This is so because a linear spring is being used to simulate the non-linear crumpling properties of sheet metal.

Mr Curtis, Mr Barley and Dr Mills should be congratulated on their attempt to introduce theory into crash testing, this is probably the most difficult structural problem still to be solved in the motor industry. Finding a satisfactory solution to the problem will be a long, but rewarding, process because, in the end, it will save design and development time.

I would like to ask Mr Curtis two questions about side impact: can he say from a theoretical study what the effect of increasing the overall beam strength would be, and what should be the beam strength of the door?

For a 45° impact a balance has to be struck between penetration and the lateral acceleration of the occupant against the structure. The photographs suggest that penetration is too great on existing vehicles and may cause lethal accidents.

It is necessary to discover whether it is possible to specify a stiffness for the side of the car, including the door beams, which will give a reasonable compromise between intrusion and deceleration without interposing, say, two feet of crushable material between the occupant and the outer skin of the car.

Mr C. P. Silvester—Mr Gough said at the end of his talk that he is going to a 2 mm laminated windscreen. The American Federal Authorities stipulate laminated windscreens, the Swedish and Norwegian Governments require laminated windscreens. What action will it take for the UK to follow? Will it need legislation or will the move come from the industry?

Mr A. E. Ballard—With regard to the radiator grille, my experience is that with a plastic moulding of this size it is necessary to use fixings which will allow for a considerable amount of expansion and contraction under extremes of temperature, otherwise buckling will take place. Have you experienced any trouble with this?

The Chairman—I would like to ask Mr Gurney a question. Do you think, as a fabricator, that the body design technology is keeping pace with the advent of the numerous plastic materials? In other words, are we taking advantage of all the materials we can?

Mr A. A. Campbell—You mentioned the Honda ABS panel, really the only major panel you have mentioned for volume production. Can you relate the cycle time for a panel in ABS to that of one car per minute? And does the panel come coloured out of the mould? If it requires to be painted after moulding, can it withstand the oven temperatures of the normal car body paint system?

Mr D. B. Dowling—Why is it that in the last two or three years, I have driven cars in which the window winder turns the opposite way to those that have been in use for the last 40 years?

Mr C. J. Cooke—Some years ago, when I was analysing car bodies by methods very similar to those just described by Mr Tidbury, I ventured to suggest at an Institution meeting that the finite-element method would prove too costly for optimization of a car body and I suggested then that we should try to understand how the car body worked, in order to reduce the amount of computer time involved. For my pains I was called a Luddite.

Today, we have had Mr Wardill's paper which has, I think, done just this, and in Dr Moore's paper we find a graph of computer time against the number of nodes which reaches to two hours per run. I think a third-generation computer costs about £150 per hour, so that is £300 per run, which I think is still too expensive to permit extensive optimization of the structure.

Could Dr Moore advise as to whether he sees the future as incorporating more of Mr Wardill's type of work, or will the cost of the computer-run on the finite-element method come down drastically in the future?

If I could go on to more specific points in relation to Dr Moore's Fig. 7.7, I am a bit worried about the lower 'A' post. It changes its bending moment over its length by about 900 lbf in, yet it has no shear forces. How does a member change its bending moment without shear forces?

Going on to the sill bending diagram, from the end shear loads and bending moments, the sill seems to be supporting 160 lbf, while the imposed central load was 1600 lbf. Does this mean the floor panel is supporting 90 per cent of the shear force?

Finally, is it adequate to quote the member end load as a mean load? (I think that is what Dr Moore has done.) I would have preferred to have the load at each end of the member round an aperture. The member along the edge is collecting end load in terms of shear per inch, so that for the body described, where the shear per inch is about 10 lbf/in at least, the end load over the 20 in length of cant-rail would change by 200 lbf. Is the mean load sufficient in these circumstances?

Mr N. H. Watts—I would like to ask Mr Wardill to what extent his philosophy of simplification caters for dynamic problems.

Mr G. B. Bolland—Both Dr Moore and Mr Wardill have shown us examples of some pretty atrocious structural connections between sill and cross-member which have obviously been designed with production in mind. I wonder if either of them has had any influence on design, as a result of their analyses? Some slight modifications of the joints could increase the stiffness of the complete structure considerably. Could either of them say whether he has had any success in this direction, perhaps at the expense of a slight increase in production costs?

Mr A. A. Campbell—Stiffness has been mentioned. British Leyland front-wheel drive cars have stiff bodies and other bodies currently manufactured are more flexible. Is stiffness a virtue, and if so, why? Is this in fact a design criterion we are striving for by means of the structural analyses?

Dr B. B. Hundy—Certainly from the ride point of view, you can get away with bodies that are considerably less stiff than the Minis and 1800s. Plenty of experimental and production bodies have been made with very much lower stiffness which were perfectly adequate on the road. As long as door clearances are satisfactory and the windscreen does not pop out stiffness can be a rather over-rated factor.

Mr J. A. B. Wolfe—My question is really aimed at Dr Moore, but I think it embraces the whole philosophy behind Mr Wardill's paper. When our body engineers start work on an entirely new body, they move very, very quickly. Unless the structural engineer can get off the ground rapidly, any contribution he can make to the design of the new body generally comes too late. In such circumstances one is forced to use the approach of Mr Wardill. Generally, one has second thoughts about that later but when one wants to change something quite drastically, one cannot do it.

I think both Mr Wardill and Mr Cooke have missed the point about computers which are getting faster and what costs you £150 now will probably cost you £10 in ten years' time. They are here to stay. Let us make use of them.

My question to Dr Moore is this: assuming we use your approach (and I am right behind you), the problem is to increase its speed so that it becomes useful and can really affect the prototype vehicle design in the early stages. Have you any comments as to how this can be done?

The second question is, have you influenced the design of any new vehicles with your work so far? My question is really based on the fact that it takes an awful lot of man hours to prepare the data for a very complex idealization. We can do it in about five weeks if we are lucky but body engineers cannot wait five weeks. They usually cannot wait five hours. If you cannot tell them the answer in a day or so, they make their own decisions. What I am getting at is, have you any ideas on the way in which the speed of really complicated calculations can be improved?

I am not suggesting that you are completely wrong, but you can get a better answer if you do the job the right way. By the present method it takes too long and when you have got the right answer, the answer you are happy with, you have to change something. I think there are ways of improving the method.

The big structural programme is a very useful tool for designing under floor elements and so on.

Mr D. Davy—It is merely a matter of the way in which the programme is written. It is not difficult to organize the programme so that, after a complete analysis, all of the results are stored for later, selective interrogation. Thus, if only load paths are required initially, these may be obtained without automatically getting other, perhaps confusing, data.

The problem is really one of initial planning.

Mr N. H. Watts—On the dynamic problem, it seems to me unnecessary to solve the static strain problems. You have to solve dynamic problems first, and this question of whether any structure should be stiffer or softer cannot be answered until you know what the dyanmic behaviour of the structure is.

Dr Mills—In the Department of Mechanical Engineering at Birmingham, we have concentrated on dynamic and static deflections more or less in parallel. We had hoped that a dynamic problem with zero frequencies would be the same as a static problem but we have moved away from this idea and are now developing separate programmes. We have in the pipeline a complete analysis of a chassis frame—statics, stresses and dynamics. This should be appearing shortly* and will give you some idea of what our interests are. This is in fact only a beam structure problem, it does not concern plates and beams, but we are also working in that area.

Mr C. C. Norville—Mr. Tidbury said that they were concentrating on the correlation of static tests with predictions. I have analysed the underbody and the complete body of a vehicle and found the beam and torsional stiff-

* Reference: I.Mech.E. Proc. 1970–71, Volume 185 44/71, pp. 665–690.

nesses had increased five times when you added the roof of the structure. Most of the papers I have seen have given a few underbody deflections and with good correlations. Is Dr Moore quite happy that this adequately explores the correlation problem, since small changes in slope between displacement points can give large changes in stresses?

Because of the large change of stiffness due to adding the roof, the connections between underbody and roof must be exceptionally important and perhaps warrant investigation of stresses in these pillar areas.

Dr G. M. Mackay—Mr Simmonds talks about the use of controls in an emergency and has set up a procedure for determining how much reduction in impact speed you would get with quicker operation of controls. Has he any information to suggest that drivers do in fact use controls in an emergency? Has he found a set of priorities for different controls? What is their importance in an emergency? It seems to me that the brake is the only thing that is really used in emergencies with any significant frequency.

Mr C. R. Ennos—Are mounting points for accelerometers included in the dummy?

I notice he is called 'Sophisticated Sam'. Has Oscar Humanus gone and do you find Sophisticated Sam better?

A factor that worried me a little is the fact that you say he is frangible. Obviously, if we use him in impact tests, he is going to break and that brings us to the cost. Can you give us an idea of how much he costs?

We know he is far from perfect but he is probably the best so far developed. I wonder, however, why we started painting lips and eyes on him, because that makes people think he is far more real than he is. Do you think it is necessary to have visual simulation, and, particularly, hair?

Does that mean we cannot purchase Sophisticated Sam as he is now? He is purely a development tool?

Mr D. B. Dowling—Mr Simmonds made the point that a hand-control is more likely to be out of reach of the tall driver since he is sitting further back. If this is so, presumably one must place controls within his reach. What is the use of his ergosphere with the different layers?

Mr R. D. Lister—Mr Simmonds mentioned the difficulties of using the new dummy, described by Mr Waller as frangible. The choice of sizes of dummies to use is even more difficult because, to cover all possibilities you need permutations of all perceptible ranges of limb and body lengths etc. in order to be quite sure that a vehicle interior design is satisfactory.

The Chairman—I am interested in what Mr Mackay said about the position of emergency controls. But the controls you want really handy are those you need to *prevent* an emergency. You stamp on the footbrake when the emergency is right on you. It is my strong opinion that there are certain controls which must be handy to the driver and which he must be able to operate instantly and without hesitation by day or night. These are the headlamp dip, the direction indicators and windscreen wipers.

All these can be mounted on the steering column, as has been done on some recent models, and it is a practice which will be increasingly followed. There they are handy whatever the size of the driver.

I would like to ask Mr Fallis if he has investigated how many controls you can put in this position. Obviously, if you have too many, the driver may be confused, but if you keep them to the essential minimum, they are very handy on the steering column, and I feel that is where they should be.

Mr R. W. Mellor—I would very much like to know whether Mr Jacobson has evidence of early damage. I have not seen any published work on this, and if he does have any such data, I think it would be most helpful if these could be published. He is suggesting that damage occurs within the first few weeks of ownership.

Mr N. H. Watts—With regard to Mr Waller's manikin, we have to match the behaviour of a real body. I wonder whether some sort of correlation with the dynamic behaviour could be obtained with accelerometers attached to a human body. Head resonance would be a good one and abdominal viscera as well.

We have to match the elastic and dynamic behaviour of various parts.

Dr G. M. Mackay—I agree with what Jack Waller has said. There is a lot of variation.

Mr T. Karen—I wonder if Mr Fallis has done any research on colour for the inside environment? Are there some colours that are better and others one should avoid?

Secondly, is there—and if so why—a gap between having intentions and carrying them out? For example, you obviously want to provide good seats to make the driver comfortable and yet I drove a *Cortina* for a number of years and liked it very much, but if you push the seats right back to accommodate a tall person they are very bad. There are worse seats in the industry but, in this day and age, with all the research you are doing, can they not make a seat in which one is comfortable if one is nearly 6 ft tall?

If you are nearly 6 ft tall, you should be able to have your back reclining at the same angle as if you are 5 ft 3in.

Dr D. B. Rees—Has any work been done on relating response time for operating controls to the sitting attitude?

I am thinking of the truck driver's vertical position, as against the more horizontal sports-car driver's position.

Mr S. T. Balzer—As a designer, do you feel that body engineers generally are unreasonably impeding your efforts towards achieving safety?

Mr A. A. Campbell—My opinion is poles apart from that of the previous speaker. Personally, I would like to see complete standardization of controls. Can this be contained within styling requirements and would you be prepared to carry standardization as far as a standard pattern of switch layouts?

Mr M. A. I. Jacobson—I would like to congratulate the speakers on their practical approach to the problem. I want to ask Mr Greenaway, in particular, how he proposes to deal with three very practical problems which confront a good many manufacturers these days. One is adequate stiffness of jacking points.

Another is towing attachments. We are concerned here with the problem of a vehicle which is immobile for one reason or another. It may be just a straightforward tow or a pick up and tow which is not always the gentlest of operations.

The third point is the junction of, particularly, the 'B' post and its spot-welding at top and bottom, which were mentioned earlier. We find there is now a considerable problem building up on vehicles, admittedly three or four years old, of corrosion and salt spray. It appears to reduce the stiffness of the vehicle to the extent that we have sagging doors on vehicles that have been in service for some considerable time.

Mr N. H. Watts—I have two questions for Mr Rodger. First of all, does he assume that the response to his test is linear with respect to force? In other words, if he varies his force at the point of application then would the mode of vibration still be the same? Does he assume the mode only changes in amplitude?

Secondly, he seems to be concerned about the contributions to the mode of vibration from more than one point. Does not this make a case for putting in a random signal at the point of application, say white noise, and then doing the analysis at other points?

Mr C. H. Ward—I see from Mr Chappuis' results that there is very little data below 100 Hz. Has he any comments on the results of the effects of damping, insulation or absorption techniques at the lower frequencies ($<$100 Hz)? Have these methods any useful effect on noise reduction below 100 Hz?

Has Mr Rodger any comments on the effect of body bending and torsional modes of vibration on the noise levels in the body cavity? Has he found that the noise levels are higher when the body vibration mode is primarily in bending? With regard to the effects of the total test time, how does this affect the body being used for continual testing?

Mr W. E. Shirley—Could I ask Mr Rodger to enlarge on why he needs to use more than one vibrator, in view of the fact that he is using the Kennedy and Pancu method? How does he know where to put in these extra vibrations? What phase does he set them up in?

Mr C. J. Cooke—I got the impression from Mr Rodger's graphs that the structure underneath the floor is exciting the floor and I imagine there will be a lot of energy which may make damping difficult. Would it be a good thing to separate the floor from the under-floor structure so that it is not excited by it? But would this in effect form a 'thunder sheet', such as those flapped in the wings of the theatre? May I ask Mr Chappuis if it is possible to damp such a thunder sheet?

Mr M. A. I. Jacobson—At least one European manufacturer of some size does in fact take his car bodies in after they have been on the road for three or four years, i.e. just before there would be a model change, and subjects them to both the types of testing which Mr Greenaway and Mr Rodger have described. Quite remarkable differences were found although all bodies are normally produced with the same press tools. Have others come across similar variations?

Mr J. A. B. Wolfe—I would like to answer Mr Osmond's question about the large number of dial gauges on the sill. It is not a standard procedure and it is not done on every body we test. The *Avenger* sill has a high inclination of the principal axis—about 35°—and it is also a very shallow, thin sill. The early bending tests showed some peculiar discontinuities in that area. Before you can do anything about them, you have to find out exactly what they are. That is what is being done.

The Chairman—Mr Rodger has been advocating the use of multi-point excitation. Can he tell us how the costs, both in manpower and equipment, would compare with the use of single-point excitation?

Mr G. H. Tidbury—Mr Greenaway has shown two methods of loading for bending tests, shot-bags and a 'fir tree' system. To cover the case where high loads are required and where the ratio of these loads has to be varied to allow different loading conditions, Cranfield has used a system of hydraulic jacks pulling down at specified node points on the floor of the structure.

This method is illustrated in Fig. D1, and can be seen by the delegates in the laboratory of the Automobile School.

Mr S. T. Balzer—Mr Davy described the process of reading off the clay model, fairing out the lines and producing a master lines draft. Is he able to comment on the time saving, if any, and the manpower saving over conventional methods using this process?

Fig. D1. Where high loads are required in varying loading conditions this simulator rig can pull down specified nodes of the structure by means of hydraulic jacks

Mr D. W. Davy—The lines fairing programme is still in the development stage. We have senior layout men working on it but we have not yet gained sufficient experience to comment on time and manpower savings.

Mr J. Curtis—I want to ask Mr Davy about the technique of generating points which have been sampled by the machine. What sort of pitch does he sample at and do the first lines generated pass through the points or are they optimized already in the first stage?

Mr D. W. Davy—We do not sample points at a fixed pitch. The pitch varies with the form of the surface, more points being taken in high curvature areas than in flat areas.

In the first stage the layout man has the freedom to decide whether the lines pass through the input points, or are faired. Our experience has shown, however, that it is prudent to make the initial lines pass through the points. This provides a starting point from which decisions can be made about the approach to fairing the surfaces.

Mr J. Curtis—It might be a conflict between smoothing so much that you lose the shape that was intended and, if you do not smooth, getting nobbles due to an inaccuracy of the model in the first place. Is this done by judgement?

Mr D. W. Davy—Essentially yes. A major part of the problem of surface proving is a compromise between these conflicting problems.

Mr J. Curtis—You are able to move points when you have a mathematical representation of the surface, to see which suits the surface better. It is a matter of trial and error.

Mr D. W. Davy—It is necessarily an iterative process. The layout man is using his judgement to decide where he will relax constraints and allow the surface to move away from the stylists model; the computer is providing him with the results of his decisions. In this way we produce a man–computer partnership which exploits the best qualities of both the man and machine.

Mr E. Hellriegel—Does one train development engineers in working with the display facilities before you select them as development engineers?

No! Such facilities do not yet exist.

Our development engineers are being selected after several years' experience as detailers and draughtsmen. Beyond that, theoretical education and the ability to work creatively are the criteria for selection.

Mr N. K. Benson—I endorse Mr Davy's comment. It is true that computer costs are coming down and, once the information has been got into the computer, the cost of each computer run is smaller.

The second point is that, whereas the design engineer would certainly like to know immediately what he should do at the layout stage, elaborate computer calculations taking a month or two are not useless. It is better to discover then that something is wrong, rather than 18 months later. This is the philosophy behind computer calculations at the present time.

Mr A. C. Hill—In this drive for change, do you value market research applied to styling, or endeavour to predict what the customer will want and, if necessary, are prepared to educate him to accept it?

Mr C. P. Silvester—I appreciate that CADANCE is in its infancy but how does the stylist know feasibility of the shape, the formability which makes it possible to manufacture it?

It may be all right as far as the functional aspect of the car is concerned, but if you have a certain die angle or the form must be of a certain type, the system must rely on a tremendous back-up of data available.

Mr D. W. Davy—All the stylists I have ever talked to have been much against using rapid display and computers for the styling. They are happy about the concept of packaging but are unhappy with trying to style sculptured surfaces, using a two-dimensional screen, even though they can, if they wish, see it in perspective, because they say that the perception is not there. They like to work on three-dimensional surfaces and look at them full size and in proportion to one another.

Are you really, in your Company, going towards this method full tilt, or do you still have reservations about how much you can do with a computer?

Mr K. A. Osborne—With the tremendous resources of General Motors, particularly in styling studios, what is the type of personnel intake? Are they coming in from schools training stylists or has the Company a training scheme of its own for promising youngsters?

Dr D. B. Rees—To what extent do you think the large manufacturer actually dictates styling from the customer-acceptance point of view, by market saturation with a certain style?

People may become so conditioned to the shape of a certain make of vehicle produced in quantity by a large manufacturer, that the manufacturer in fact dictates what they will accept at a later date.

Mr A. E. Ballard—We hear a lot these days of steel bumpers being replaced by self skinning foam on metal. Do you think there is any future in this form of bumper?

Mr McLaren—Would you agree, that in building design constraints into a computer/plotter programme for the stylist and body design engineer, there is a danger that innovation and totally new concepts, such as Issigonis' tranverse engine layout, become blunted or, worse, never take root?

If automobile makers all build with presumably similar constraints, then the products will begin to inherit a natural uniformity, until a manufacturer ignoring the rules hits the jackpot!

Mr C. J. Cooke—When around 1956 it became obvious that any frame which could be squeezed under the next model would be too slender to affect stiffness, we decided to delete it. I think, however, that we might have made more effort to justify the existence of a slender frame, had we been in the position of having to clothe the same basic mechanicals with a new bodyshell each year.

I understand that the perimeter frame makes some contribution to the suspension of the vehicle by virtue of acting as a very stiff spring. But it has no damper: has this ever led to a problem of vibration interaction between frame and suspension?

As Mr Sherman has said, it is sometimes reasonable to allocate a zero torsional stiffness to a frame. Fig. D2 shows a crude frame which has very little inherent torsional stiffness.

Very light finger pressure applied in the direction of arrows Y_1 and Y_2, with reactions at X_1 and X_2, distorts it easily. Let us imagine members *ab* and *cd* to be longitudinal torsion bars and members *de* and *fb* as transverse suspension levers, supporting a wheel on the side of the car opposite to the associated torsion bar. It can be seen that, by applying the rear torsion couple via loads in the directions of Z_1 and Z_2 we can make the frame immensely stiff in torsion.

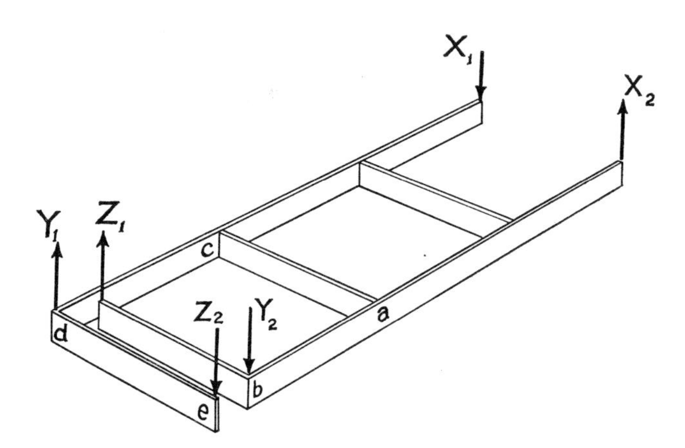

Fig. D2. The stiffness of this type of frame can be greatly improved by changing the points of load application

Prior to arriving at this geometry I would have thought it impossible to deceive a frame into behaving as if it were not in torsion, however one applies the loads, but the trick in this case is to convert the couples twisting the members into bending moments.

Mr K. A. Osborne—How much of this change to the use of chassis frames in America was brought about by the corrosion factor?

Mr M. A. I. Jacobson—I wonder whether the legislative forces in the United States have had something to do with that as well, as regards frontal impact and side impact intrusion? Is this type of automobile safer because of its frame construction?

The Chairman—It is untrue to say that we do not use chassis frames in this country. There are a lot of cars running about with a front frame and a rear frame which carries the suspension and rear axle, which is exactly what American frames are doing. However, we cut out the two side-rails, which seem to be doing nothing at all. Why does Mr Sherman put them in?

Mr W. E. Shirley—Can you tell us what signal you use for controlling your hydraulic shakers?

What is the cycle time of your impulses?

Mr Sherman—About 15 minutes and it plays over and over again, as long as we wish.

Mr Dunn—If I might highlight this result that Mr Phillips mentioned, the absolute noise level in a given vibrational mode inside a structure is controlled by the volume velocity of the structure in that mode. There is no direct linear relationship between the point mobility measured when that mode is being excited and the resultant noise level. I would not expect there to be one. The reason why any one structure in a given mode generates a lot more noise than another one is because of the volume velocity factor. This is the control on the thing. You may only get, say, 5 or 6 dB difference in driving point mobility for two similar modes and two different structures, but the volume velocity for these respective modes may be totally different, which will change the generated sound pressure by quite a large amount.

I do not know whether those of you who are interested have scanned available mobility plots for various structures. Mr Phillips was talking about the difference between two nominally identical structures. In fact, if you do tests on a saloon car and similar tests on a van structure, the mobility levels lie within a 20 dB band. You have got to change the structure rather drastically before you move the overall mobility response level significantly up or down.

Our programme to date has not included any detailed study of aspect ratio effect on the dynamic performance of the tyre. The rig will certainly discover changes in the dynamic stiffness if they do exist because of its extreme sensitivity to any changes at all.

We have done some work on testing two nominally identical tyres. In each case we have found that the curves lie very close together and I would conclude that it is satisfactory for a manufacturer to test only one of a given type of tyre.

With respect to damping, the difference in resonant circle diameter is very small. This is surprising, I feel, but it is certainly the case.

On the question of repeatability between any two given tests on the same tyre, I have found that, starting from square one, the response curves, in particular the mobility phase-angle (which is perhaps the most sensitive parameter obtained from the dynamic tests), agree to within less than 0·5°. This, I am sure you will appreciate, is quite good. The absolute level of mobility certainly agrees for two controlled tests to within less than 0·5 dB.

The Chairman—This is the end of the symposium and it is obvious that a tremendous amount of work has gone into it. On your behalf I would like to thank the Organizing Committee for all the hard work they have done.

Authors' Replies

C. R. Ennos—In reply to Dr G. M. Mackay, this procedure has indeed been tried by one or two testing establishments, but I do not think that it improves the vehicle-to-vehicle side impact simulation. As soon as the vehicle is put on castors, problems are encountered with unrepresentative vehicle height, wheel-loading, and the post-impact motion of the test vehicle. A case could be made for it to simulate sliding impact with a fixed object, but here again a suitably shaped mobile barrier impacting on the stationary vehicle at the correct angle is probably a better and simpler approach.

In reply to Mr G. H. Tidbury, I do not believe that it is possible to define in absolute terms an overall stiffness of the vehicle side. The suggestion with regard to maintaining structural continuity, for a 50 km/h, 45° impact between similar vehicles, made in the final paragraph of my paper, is one approach to resolving the problem. An interesting corollary to providing two feet of padding is that the attendant increase in the vehicle width could provide a hazard on our narrow European roads which increased, rather than decreased, the total number of injuries sustained in road accidents.

In reply to Mr R. D. Lister, I did not intend to imply that intrusion was always bad, obviously work done in deformation can be beneficial to the vehicle occupants. It absorbs some of the bullet vehicle's kinetic energy, which would otherwise go into increased acceleration of the target vehicle, thus increasing the severity of the secondary impact between the occupant and the interior structure (see Fig. 1.15).

The matter is not simple, however, as the increased stiffness of the target vehicle's side structure may induce both vehicles to rotate, and thereby reduce the severity of the secondary occupant-to-interior impact. It is not correct, therefore, simply to relate structural stiffness with the one in five injuries due to crushing, as it can also influence the more general case of injuries sustained from secondary impact with the interior. Any answer will have to be a compromise, and it is with this in mind that the recommendation in the final paragraph of my paper was made.

Making the comparison between vehicle-to-vehicle and frontal barrier impacts on a velocity-change basis does not overcome the problem I was trying to illustrate. With dissimilar vehicles, the differing locations of the hard and soft points of the two structures result in deformation patterns totally different to those of the barrier or identical vehicle-to-vehicle impacts. The deceleration/time relationship will consequently also be different, and can drastically affect occupant kinematics.

In reply to the question by Mr W. H. Watts, the deceleration of the vehicle is normally recorded at several different points on the body where structural deformation will not occur. A standard position used in most tests is the base of the 'B' post. There are a variety of accelerometers that can be used to record the deceleration transient. At Ford we favour the small piezo-resistive type, fitted to rigid, tri-axial mounting blocks.

With regard to filtering, perhaps I should re-emphasize that we filtered down to 20 Hz purely to investigate the possibility of characteristic pulse shapes existing for specific types of vehicle. Clearly, at this filter level, which would require 25 ms as we are concerned with a half sine pulse, detailed measurements would not be undertaken.

Finally, in reply to Mr A. C. Hill, one of the major causes of fatalities in road accidents is ejection, and for this reason considerable emphasis is being placed on retaining the occupant in the vehicle during impact. A bonded-in windscreen, utilizing glass with a known impact performance, can form an essential part of an occupant restraint system. If a screen is used which is likely to shatter on impact, then it may be advantageous to allow it to 'pop out' under inertia loading. In this case, however, one would have to ensure that the exposed glass retention system did not constitute a hazard.

Mr D. W. Sherman—Mr C. J. Cooke's frame demonstration on torsional rigidity is entirely correct. It is quite understandable. Such resistance in a complex structure depends upon the resistance of the members to both bending and torsion. The American box rail frame is based upon this principle, the rails resisting by torsion and bending, and the cross members serving essentially as beams, acting as levers to twist the rails. Adherence to this principle reduces the labour needed to make the frame and therefore achieves maximum torque resistance at least cost. However, the cross-member type frame (pure beam) is more efficient and should be used when space is available.

In reply to Mr M. A. I. Jacobson, the frame in today's American cars is used almost entirely because of its isolation effect, reducing both noise and the forces acting on the structure. Also, it softens the car and gives an easier ride. However, we believe it is also better from a collision standpoint, particularly with respect to potential improvements.

In reply to the Chairman, with the very soft rubber frame–body connectors used on American cars the centre rails serve to prevent relative movement between front and rear. While, without the rails, such movement is minute, it does influence the feeling of the car on the road. Also, while the rail section is small and cannot, therefore, have much structural influence, it does not weigh much either. In fact, the effect/weight ratio is good. In addition, the full frame permits chassis building in the open, ahead of body installation.

In reply to Mr K. A. Osborne, corrosion is not a problem with metal frame thicknesses, and we feel this to be an additional advantage of frame-type construction.

In reply to Mr W. E. Shirley, the shakers are controlled by magnetic tape on which has been recorded the vehicle's wheel action as it passed over selected road surfaces. Displacement, acceleration and velocity are all measured. The shakers then force the axles to duplicate the road action.

However, this is all purely vertical and lateral forces arising from vertical wheel action are most important. Since the lateral movement responsible for these forces is too small to be measured as described, the resulting strain in key members is measured instead. Then, on the test rig, lateral link ties are calibrated to produce the same strain, as a function of vertical wheel (axle) displacement. The result is a faithful duplication of road action. The shaker permits structural action and distortion to be observed.

Tests are accelerated by constant testing round the clock on the types of roads the car will have to run on.

As to cycle time, the tape runs for 15 minutes and two duplicates are used. When one is complete, the other switches on automatically so that operation is continuous.

Mr G. Simmonds—In reply to Mr D. B. Dowling, the main point about the 'ergosphere' approach is that it is not necessary to assume that a driver is sitting in any particular position or that he has any particular dimensions. The technique is to set a representative group of people in a vehicle package and *measure* their reach at specified vertical and lateral positions. Thus, at any of these specified positions it is possible to calculate percentiles of reach.

The 'layers' of the ergospheres are the surfaces joining points for the various percentiles. Thus, it is not even necessary to assume that the limit of reach is always imposed by a particular individual. In general, this will not be the case. For example, Mr A may reach further forward along a line well above his head than Miss B, but the reverse may be true along a line behind the steering wheel.

The use of the layers is simply to assist in estimating the proportion of drivers that will be able to reach a particular control. Alternatively, it may be decided that, say, 95 per cent of drivers should be accommodated; in this case only the 5th percentile 'shell' need be used.

The exact method of applying the method is still being developed for ISO.

I agree with Mr R. D. Lister that we need different sizes of dummy in order to evaluate the interior design of a vehicle. The only point I was making was that the term 'percentile' is being misused. I would prefer that we did not refer to '95th percentile manikins', etc. More accurate descriptions would be 'large man', 'small woman', 'average driver', etc.

Since we have not succeeded in quantizing the sum-total of human dimensions we are not justified in referring to '95th percentile man'. Even if we restrict ourselves to lower leg, upper leg and torso-plus-head lengths, the fallacy is clear. The sum of these 95th percentile lengths is about two inches longer than the 95th percentile standing height. This is quite simply because people with, for example, long thighs tend not to have quite such extreme lengths in the other bones. Thus, the so-called '95th percentile' dummies have, in fact, sets of arbitrary dimensions which are adjusted to give 95th percentile stature and weight.

There is nothing wrong in making such tools, in fact they are useful. But we would do well to realize that '95th percentile dummy' is a name which has little more meaning than 'Big Joe'.

I quite agree with Dr G. M. Mackay that the main control used immediately prior to an accident is that for the brake. However, I believe that there are other critical situations where my remarks apply. Indeed, one of Dr Mackay's colleagues, Dr Clayton, records an accident which illustrates the general point.

An accident situation developed as a driver was searching for a control. While this does not answer the direct point of the question, it does point to the more general need for higher compatibility between car and driver. What we certainly need is far more objective knowledge of how a driver behaves, both immediately prior to an accident and in the longer, critical phase that precedes the time when a crash has become inevitable.

I agree with the Chairman that there is probably a small number of controls which are of particular importance.

Mr G. O. Gurney—In reply to the question by the Chairman: On the question of plastics, design technology for body engineering is, of course, primarily related to sheet steel at this time. The fundamentals are the same, however, whether designing for steel or plastics. Where the difficulties occur is that plastics have differing characteristics, some of a more limiting nature than steel but others allowing far greater freedom.

The motor industry is not taking *full* advantage of new materials and this is probably a question of confidence. Basically, this is an education problem which is being tackled, albeit somewhat slowly.

When plastics are being used to replace metals, the are expected not only to give economic advantages with the same performance but they are often expected to overcome the deficiencies of the metal part. The fact that this is often achieved sometimes goes unnoticed.

However, there has been considerable evolutionary, rather than revolutionary, progress in the introduction of plastics which shows that there are many body engineers who are fully aware of the properties of these comparatively new materials, or who seek out the information they require.

In reply to Mr A. A. Campbell, I am not quite sure what is meant by the question about the Honda ABS. Very few car manufacturing plants have achieved a production rate of one per minute and those that have are not fully able to sustain this for long periods or to work the track round the clock.

The average moulding cycle time for major panels is certainly in the order of one per minute and injection moulding machines are regularly run round the clock. I cannot see capacity limitations, therefore.

Regarding paint systems, I would first question the premise that there is a normal car body paint system. There are many systems and the temperatures vary considerably. It is true that the latest systems for high output of steel bodies are operating at increasingly high temperatures which could cause embarrassment to major panels manufactured from certain thermoplastics, such as ABS.

The spectrum of plastics, however, is large and, if a manufacturer needs to go to high-temperature paint systems, then he would use other materials, possibly the reinforced thermosetting plastics.

To the specific question, 'does the Honda panel come coloured out of the mould?' the answer is yes, but it is nonetheless painted with the normal top coat finish used for the rest of the body to give similar fade characteristics as on the metal parts.

Mr M. Rodger—With regard to Mr N. H. Watts' second question, the use of a random signal really only complicates the problem because the body will respond in some degree to all the frequencies of excitation. It then becomes a difficult analysis problem to extract the body resonance-characteristics for any individual resonance. The approach put forward in my paper is to simplify the analysis by controlling the inputs to achieve the purest resonant modes possible.

In reply to Mr C. H. Ward, as a general rule bending modes are more effective noise generators than torsional ones, but as with all the best rules there are many glaring exceptions. It is these exceptions that helped to set off the investigations into the relationship between interior noise and body motion which are referred to in the paper.

The duration of the test does have an effect on body response but, at the force levels used, the amount of 'drift' is fairly small. Variations of a similar order do occur from day to day, probably due to changes in ambient levels.

Considering the differences which can exist between similar bodies, it is not at present worth while to try and pin down these daily variations any closer.

In reply to Mr W. E. Shirley, the Kennedy and Pancu method enables one to detect the presence of modal interaction at resonance. The use of multipoint excitation enables this interaction to be reduced, as explained in the section 'Multipoint excitation'. The procedure for finding input points is an iterative one, similar to that used in single point work. As explained in the paper, the phase of the master exciter is servo-controlled to hold a resonant force–velocity relationship. The phase of the other vibrators is controlled indirectly by adjusting the force levels until they too have correct resonant relationships.

In reply to Mr C. J. Cooke, it is certainly true that, in the resonances which I have illustrated, the whole of the structure of the body is participating, and it would be very difficult to damp. The separation of floor from understructure rather implies a separate chassis or sub-frame, which does offer further possibilities for isolating the body from vibration inputs.

In reply to the Chairman's question, no direct cost comparison is possible since the uses of single and multipoint excitation are complementary rather than competitive. I would not recommend using any form of multipoint equipment without first making extensive use of single point.

In our case, a five-channel, multipoint system is used in conjunction with a fairly sophisticated single-point system, and the multipoint facility represented an extra cost of about 40 per cent.

Mr A. Chappuis—In reply to the question by Mr C. H. Ward, noise at lower than 100 Hz is usually emitted by large surfaces, in this case by a large part of the body, including parts of the structure.

The efficiency of a local surface-damping treatment is low. However, new techniques, still in the experimental stage, should lead to damping of the structure with positive results.

Sound-insulation of iron sheets for very low frequencies is very difficult to realize, especially when these panels are excited by structural vibration.

The effect of all kinds of sound insulating material depend mainly on their weight.

The use of an absorbent porous layer is illusory because for the frequencies concerned the thickness required is enormous. Techniques based on acoustical resonators or membrane effect allow high absorption at very low frequencies but their efficacity is limited to a very narrow frequency band. For these techniques to be of value, it would be necessary to have a series of elements tuned to cover the whole band of frequencies at which the sound level should be reduced.

In general, the efficiency of soundproofing materials is much reduced at very low frequencies. Therefore it is advisable to try to reduce noise, and especially vibration, at the sources in this frequency range and search for the optimum mechanical mounting of the sources on the body in order to decrease the vibratory energy transfer and to avoid bending and torsion resonances of the structure.

Mr J. Fallis—In reply to questions by Mr T. Karen: naturally, we have done a considerable amount of work on colour for inside environment. It would seem that the public themselves have made a choice, surprisingly similar to our findings. Colour perception by individuals differs greatly and is biased to some extent by the introvert/extrovert character of the subject. In all events, an overpowering and garish scheme tends to create nervousness and I am sure that no one is surprised at this. Again, the level of garishness acceptable is inclined to depend on the subject. I do feel, however, that our method of measurement leaves much to be desired and we would welcome a more sophisticated, meaningful way to get results.

Secondly, I think, and have said, that there has been a marked improvement in seating. The latest 'Cortina' has, without a doubt, better seats than the previous model mentioned in the question.

Replying to Dr D. B. Rees, there has been a lot of work done on this question and is better answered by the ergonomics expert, for instance Mr G. Simmonds. However, I believe that seating attitude, reach/fatigue have a very definite influence on response, and it is a subject in which the interior designers are involved.

I do not believe that a vertical driver position, as against the traditional sports car position, makes a lot of difference to response provided both postures are designed to cause the least amount of fatigue.

In reply to Mr S. T. Balzer, I believe that body engineers are inclined to drag their heels over any subject that they have not experienced before. Whilst safety, particularly environmental safety, is not something we have just discovered, it has recently achieved more notoriety and, therefore, has become a new subject to the engineer.

I recognize the anxiety of engineers not to launch into foolhardy schemes but to tread the sensible path which, after all, is what the designer wants too. We feel that we have to use our discretion to pursue a sensible approach to a safety environment and we invite the engineer to join us in this.

In reply to a question by Mr A. A. Campbell, I think we are probably saying the same thing, inasmuch as standardization of controls, pattern and graphics is quite within the limits of possibility. The reason for standardization is to reduce confusion to the driving public. I had hoped that I had made it clear that I applaud the idea of standardization, certainly of graphics and all switch patterns. Certain controls must be grouped on either side of the steering wheel if they are needed for emergency action, whereas other controls, for instance radio and heater, should be made available to both driver and passenger. This is done by most motor manufacturers but it does need tightening up. We are all working towards standardization via the ISO. Even within its limits there is still room for a car's character to show through. The graphics, however, are still an international mess and I would very much like to see a great improvement immediately.

Mr C. M. MacKichan—In reply to Mr A. D. Hill, we place some value upon market research but sincerely believe that most results we have seen can tell you where you have been, but not where you should go. Historically, the artist or designer has always been a bit ahead of his society. We attempt to stay ahead of our consumers and so far, I feel, we have. It is not a clear-cut, black-and-white situation, but we use less formal market research than many would imagine, for its worth has not been decisively demonstrated.

In connection with a question by Mr C. P. Silvester, we don't intend the CADANCE system to include the complete definition of specific mechanical components. We expect it to decrease dramatically the time required to establish parameters for a given direction of a design. It will be a very effective tool for the designer. With reference to the example you cite, the bumper, feasibility of manufacture is not thought of as being part of the computer output. Dimensional parameters, amount of spatial volume required, distance and direction of stroke (if it is energy-absorbing), these are obtainable.

The third comment is quite correct. These problems are worked out between the designer and the engineer from the car producing division in a co-operative effort.

In reply to comments by Mr D. W. Davy, we realize the anti-computer bias of many designers and admit that it has some merit. We see the computer as a tool for dynamic simulation of door swings, compartment opening, passenger displacement upon impact, etc. We have never felt that the car could be surfaced, in a creative sense, completely in the computer or evaluated in two dimensions. However, a three-dimensional, holographic simulation could be another matter. This as yet is an unreached frontier. We also see the computer as a great aid in allowing us to sculpture full-size models nearly automatically.

In reply to Mr A. E. Ballard, we pioneered the use of skinned micro-cellular urethane foam bumpers on the Pontiac GTO of 1968 and currently they are featured on the GTO, Pontiac *Firebird*, and Chevrolet *Camaro* Rally Sport. Our material is not self-skinning, however.

With regard to comments by Mr McLaren, we see computer-assisted design as enabling a designer to assemble his own set of parameters based upon the spatial requirements, both static and dynamic (i.e. wheel-jounce, rebound and turn envelope), rapidly and in a variety of combinations. Where it now takes time-consuming labour to test the results of 'thinking a way around these parameters', the system we visualize will let him verify his insights, or intuition if you wish, almost instantaneously. I'm sure Alec Issigonis would have welcomed such a versatile tool—as we will when it is finally developed. Far from creating uniformity, the number of possibilities may become mind-boggling, requiring a high level of judicial thinking for a final selection. I see no danger to innovation in the use of the computer.

All U.S. manufacturers must build to rules established by the government in *certain* areas, or face stiff fines. No manufacturer can afford to ignore *these* rules. We believe doing effective design within their constraints to be a worthy challenge to our skills and, in many cases, expect to exploit them by turning them into virtues.

With regard to Dr D. B. Rees's question I know of no measurement of the extent to which the large manufacturer dictates style to the consumer by market saturation of a certain style. In our country we have a great 'follow-my-leader' game and so the market does get saturated rapidly. We like to think we make the boldest moves at GM—at least we have recently. Because we don't design directly on the basis of consumer data, I suppose you could say we influence style. But if the customer doesn't like it, he doesn't buy, as in the case of our 1966–1967 *Tornado*. That was too bold too soon. We certainly didn't *dictate* style with that car!

List of Delegates

Adcock, B. A.	Vauxhall Motors, Luton, Beds.
Alderman, W.	David Brown Tractors Ltd, Huddersfield, Yorks.
Ancliff, A.	Rover Co. Ltd, Solihull, Warwickshire
Anderson, D. T.	University of Birmingham, Birmingham
Ashton, S. J.	University of Birmingham, Birmingham
Bailey, P.	Ogle Design Ltd, Letchworth, Herts.
Ballard, A. E.	Standard-Triumph Motor Co. Ltd, Coventry, Warwickshire
Balzer, S. T.	Ford Motor Co. Ltd, Basildon, Essex
Barber, H. R.	Pressed Steel Fisher Ltd, Cowley, Oxford
Barley, G. W.	Rootes Motors, Coventry, Warwickshire
Barrow, J. H. H.	Vauxhall Motors Ltd, Luton, Beds.
Batley, G. C.	Ford Motor Co. Ltd, Basildon, Essex
Bee, A. J.	Rolls-Royce Ltd, Crewe, Ches.
Bennett, A.	Pressed Steel Fisher Ltd, Cowley, Oxford
Benson, N.	Ford Motor Co. Ltd, Basildon, Essex
Bierman, G.	Marbon Europe N.V., Holland
Bolland, G. B.	Lanchester Polytechnic, Coventry, Warwickshire
Burley, A. J. F.	Ministry of Transport, London
Campbell, A. A.	Ford Motor Co. Ltd, Basildon, Essex
Campbell, G. R.	National Engineering Laboratory, East Kilbride
Chappuis, A. M.	Interkeller AG., Zurich, Switzerland
Childs, A. J.	Vauxhall Motors Ltd, Luton, Beds.
Clarke, F. W.	AC-Delco, Dunstable, Beds.
Cleminson, A.	Triplex Safety Glass Co. Ltd, London
Cooke, C. J.	Rolls-Royce Ltd, Crewe, Ches.
Cooper, F. D.	Ford Motor Co. Ltd, Laindon, Essex
Cornacchia, F.	Fiat, Turin, Italy
Curtis, J.	Vauxhall Motors Ltd, Luton, Beds.
Das, N. K.	Luton Body Dies, Luton, Beds.
Davy, D. W.	Pressed Steel Fisher Ltd, Cowley, Oxford
De Bank, D.	AC Delco, Dunstable, Beds.
Dowling, D. B.	Smiths Industries Ltd, London
Doyle, R. A.	Pressed Steel Fisher Ltd, Swindon, Wilts.
Drury, C. J.	Pressed Steel Fisher Ltd, Cowley, Oxford
Eagles, A. C.	I.C.I. Ltd, Hyde, Ches.
Emmerson, W. C.	Pressed Steel Fisher Ltd, Cowley, Oxford
Ennos, C. R.	Ford Motor Co. Ltd, Laindon, Essex
Fahenstock, W. S.	Dana Corporation, Reading, Pa, U.S.A.
Fallis, J. E.	Ford Motor Co. Ltd, Laindon, Essex
Ferriday, J.	Alcan Booth Industries, Banbury, Oxon.
Fowler, J. E.	Pressed Steel Fisher Ltd, Cowley, Oxford
Gallagher, J. G.	G.K.N. Sankey Ltd, Wolverhampton
Gijsels, L. F. J.	Van Hool and Sons, Belgium
Girling, P. H.	Vauxhall Motors Ltd, Luton, Beds.
Goodwin, I.	Rover Co. Ltd, Solihull, Warwickshire
Gough, H.	Triplex Safety Glass Co. Ltd, Birmingham
Greenaway, W. R.	Chrysler (U.K.), Coventry, Warwickshire
Greenland, R.	AC-Delco, Dunstable, Beds.
Gwinn, E. C.	Rootes Motors Ltd, Coventry, Warwickshire
Gurney, G. O.	Hills Precision Die Castings Ltd, Birmingham
Ham, P. E.	Rootes Motors Ltd, Coventry, Warwickshire
Harburn, J.	Vauxhall Motors Ltd, Luton, Beds.
Harding, B.	Ford Motor Co. Ltd, Laindon, Essex.
Harris, J.	Road Research Laboratory, Crowthorne, Berks.
Hellriegel, E.	Ford Motor Co. Ltd, Basildon, Essex.
Hemingway, N. G.	The Polytechnic, Hatfield, Herts.
Hill, A. C.	Pressed Steel Fisher Ltd, Cowley, Oxford
Hole, C. J.	Rover Co. Ltd, Solihull, Warwickshire
Holt, G. J.	Jensen Motors Ltd, West Bromwich, Staffs.
Holtum, C.	Jaguar Cars Ltd, Coventry, Warwickshire
Hopkins, N. S.	I.C.I. Ltd, Hyde, Ches.
Horwood, S.	Pressed Steel Fisher Ltd, Cowley, Oxford
Howe, J. M.	Hatfield Polytechnic, Hatfield, Herts.
Hundy, B. B.	Pressed Steel Fisher Ltd, Cowley, Oxford
Hurley, A. G.	Rootes Motors Ltd, Coventry, Warwickshire
Ireland, A. L.	Weathershields (Berkley) Ltd, Birmingham
Jackson, A.	The Polytechnic, Hatfield, Herts.
Jacobson, M. A. I.	Automobile Association, London
Jennequin, G.	I.S.M.C.M., St Ouen, France
Johnson, P. F.	University of Birmingham, Birmingham
Jones, I. D.	Lotus Cars Ltd, Norwich, Norfolk
Karen, T.	Ogle Design Ltd, Letchworth, Herts.
Kelkin, P. B.	B.L.M.C., Longbridge, Birmingham
Kimber, M. D.	The Polytechnic, Hatfield, Herts.
Knowles, J. E.	Smiths Industries Ltd, London
Koltermann, R. P.	Dana World Trade Corporation, Indiana, U.S.A.
Lacambre, M.	Renault, Rueil 92, France
Lavender, D. C.	Oldfield-Keller Ltd, London
Lawrence, R. J.	Vauxhall Motors Ltd, Luton, Beds.
Leeming, P. B.	Jaguar Cars Ltd, Coventry, Warwickshire
Le Salver, R.	I.S.M.C.M., St Ouen, France
Lister, R. D.	Road Research Laboratory, Crowthorne, Berks.
Long, A. R.	Pressed Steel Fisher Ltd, Cowley, Oxford
Long, T. C. A.	Rootes Motors Ltd, Coventry, Warwickshire
Lynch, G. T.	Duple Coachbuilders Ltd, Blackpool, Lancs.
Mackay, G. M.	University of Birmingham, Birmingham
Macmillan, R. H.	M.I.R.A., Nuneaton, Warwickshire
Maserati, A.	Vignale S.p.A., Turin, Italy
Maskew, R.	Bayer Chemicals Ltd, Richmond, Surrey
Massey, A. E.	Standard-Triumph Motor Co. Ltd, Coventry, Warwickshire
Mellor, R. W.	Ford Motor Co. Ltd, Laindon, Essex
Meyer, J.	Ford Werk AG, Cologne, West Germany
Moore, C. E.	Rover Co. Ltd, Solihull, Warwickshire
Moore, G. G.	Pressed Steel Fisher Ltd, Cowley, Oxford
Myatt, R.	Rootes Motors Ltd, Coventry, Warwickshire
Neale, E. W.	Hallam, Sleigh and Cheston Ltd, Birmingham
Neal, P. D.	Fibreglass Ltd, Birkenhead, Ches.
Neilson, I. D.	Road Research Laboratory, Crowthorne, Berks.

Newman, R. N.	Chrysler (U.K.), Coventry, Warwickshire
North, R. H.	Vauxhall Motors Ltd, Luton, Beds.
Norville, C. C.	University of Birmingham, Birmingham
Oliver, S. W.	Chrysler (U.K.), Coventry, Warwickshire
Osborne, K. A.	Pressed Steel Fisher Ltd, Cowley, Oxford
Osman, D. F. M.	Pressed Steel Fisher Ltd, Swindon, Wilts.
Osmond, E.	Standard-Triumph International, Coventry, Warwickshire
Pawlowski, J. W.	The Centre of Maintenance and Repair of Vehicles, Warsaw, Poland
Pearse, D. A. J.	Pressed Steel Fisher Ltd, Cowley, Oxford
Pearson, M. A.	Marbon U.K. Ltd, Clayton, Lancs.
Pilling, K.	Metro-Cammell Weymann Ltd, Birmingham
Piziali, A. P.	Ford Motor Co. Ltd, Basildon, Essex.
Plackett, A. G.	Smiths Industries Ltd, Witney, Oxon
Pollitt, J. M.	Kangol Magnet Ltd, Carlisle, Cumberland
Pope, D. J.	Vauxhall Motors Ltd, Luton, Beds.
Pope, J.	Marbon Europe N.V., Holland
Rayner, I. E.	Rootes Motors Ltd, Coventry, Warwickshire
Rees, D. B.	Vauxhall Motors Ltd, Luton, Beds.
Reid, J. E.	Chrysler (U.K.), Coventry, Warwickshire
Reimer, B. R.	Hayes-Dana Ltd, Thorald, Ontario, Canada
Rodger, M.	Ford Motor Co. Ltd, Laindon, Essex
Russell, M. J. C.	Scammell Lorries Ltd, Watford, Herts.
Rushton, G. D.	Rover Co. Ltd, Solihull, Warwickshire
Saunders, B. B.	Auto Body Dies Ltd, Dunstable, Beds.
Saward, R. W.	Rootes Motors Ltd, Coventry, Warwickshire
Scott, B. F. J.	British Ropes Ltd, Newcastle upon Tyne
Shelley, J. R.	Dennis Bros Ltd, Guildford, Surrey
Sherman, D. W.	Dana Corporation, Michigan, U.S.A.
Sheron, H.	Rootes Motors Ltd, Coventry, Warwickshire
Shilcof, L. A.	Joseph Lucas (Electrical) Ltd, Birmingham
Shirley, W. E.	Pressed Steel Fisher Ltd, Cowley, Oxford
Sibthorpe, A. H.	Ford Motor Co. Ltd, Basildon, Essex
Silvester, C. P.	Ford Motor Co. Ltd, Dunton, Essex
Simmonds, G. R. W.	Ford Motor Co. Ltd, Basildon, Essex
Snedker, G.	Vauxhall Motors Ltd, Luton
Spedding, C. E.	G.K.N. Group Technological Centre, Wolverhampton
Speed, S. J. A.	London Transport Executive, Chiswick, London
Spragg, D. J.	Ministry of Transport, London
Stanghan, R. M.	Vauxhall Motors Ltd, Luton, Beds.
Stock, C. A.	Ford Motor Co. Ltd, Basingstoke, Hants
Streeton, D. A.	Rootes Motors Ltd, Coventry, Warwickshire
Suthurst, G. D.	Ford Motor Co. Ltd, Laindon, Essex
Taylor, W. S.	Chrysler (U.K.) Ltd, Coventry, Warwickshire
Thaxter, D. G.	Vauxhall Motors Ltd, Luton, Beds.
Thurgate, J. C.	Alcan Booth Industries, Banbury, Oxon
Tidbury, G. H.	Advanced School of Automobile Engng, Cranfield, Beds.
Tree, D. W.	Chrysler (U.K.) Ltd, Coventry, Warwickshire
Wall, M. A.	Pressed Steel Fisher Ltd, Cowley, Oxford
Waller, J. A.	Vauxhall Motors Ltd, Luton, Beds.
Ward, C. H.	Vauxhall Motors Ltd, Luton, Beds.
Warren, J. W. L.	The Polytechnic, Hatfield, Herts.
Watts, N. H.	The University, Sheffield
White, R. D.	Pressed Steel Fisher Ltd, Cowley, Oxford
Wilks, P. M.	Rover Co. Ltd, Solihull, Warwickshire
Wilson, B. P.	Marshall of Cambridge (Eng.) Ltd, Cambridge
Wilson, P. S.	Chrysler (U.K.) Ltd, Coventry, Warwickshire
Winterbottom, O. C.	Jaguar Cars Ltd, Coventry, Warwickshire
Wisdom, R. D.	Rootes Motors Ltd, Coventry, Warwickshire
Wolfe, J. A. B.	Rootes Motors Ltd, Coventry, Warwickshire

Index to Authors and Participants

The names of authors and the number of pages on which papers begin are printed in bold type.

Ballard, A. E., 143, 148
Balzer, S. T., 146
Barley, G. W., 26
Benson, N. K., 148
Bolland, G. B., 144

Campbell, A. A., 143, 144, 146
Chappuis, A. M., 126, 152
Chilver, A. H., 142
Cooke, C. J., 143, 146, 148
Curtis, J., 19, 147

Davy, D. W., 134, 144, 147, 148
Dowling, D. B., 143, 145
Dunn, 149

Ennos, C. R., 6, 145, 150

Fallis, J. E., 84, 153

Greenaway, W. R., 102
Gurney, G. O., 34, 151

Hellriegel, E., 147
Hill, A. C., 142, 148
Hundy, B. B., 144

Jacobson, M. A. I., 146, 149

Karen, T., 145
Kay, S. E., 44

Lister, R. D., 142, 145

Mackay, G. M., 142, 145
MacKichan, C . M., 1, 153
McLaren, J., 148
Mellor, R. W., 145
Mills, B., 26, 144
Moore, G. G., 55

Norman, A. E., 50
Norville, C. C., 143, 144

Osborne, K. A., 148, 149

Palmer, G. M., 142

Rees, D. B., 145, 148
Rodger, M., 108, 152

Sherman, D. W., 87, 150
Shirley, W. E., 146, 149
Silvester, C. P., 143, 148
Simmonds, G. R. W., 96, 150

Tidbury, G. H., 71, 143, 146

Wardill, G. A., 62
Watts, N. H., 142, 143, 144, 145, 146
Ward, C. H., 146
Wolfe, J. A. B., 144, 146

Subject Index

Titles of papers are in capital letters

Absorption, noise reduction, 133, 146, 152
Accelerometer traces, impact tests, 7, 142, 150
'Air bag', collision protection, 3, 84
Air circuit and air ram, linear accelerator rig, 29
America, United States of; A.S.A.E. test rig, 68
chassis frame development, 88, 149, 150, 151
frontal impact test standards, 9
S.A.E. mobile barrier, impact tests, 14
America, United States of, safety regulations; door latches, 50
fuel tank, 9, 12, 19
steering wheel, 9, 19
windscreen, 9, 19, 143
Anthropometrics, 85, 86, 97
Aperture distortion measurement, 105
AUTOMOBILE BODY TESTING TECHNIQUES, 102
AUTOMOTIVE DESIGN WITH SPECIAL CONSIDERATION FOR SAFETY IN INTERIOR DESIGN, 84

'B' post; corrosion, 146
strength, 15, 146
Barrier impact simulation programme, 20, 143
Barrier, S.A.E. mobile, impact tests, 14
Bending loads, buses, 75, 81, 82
Bending tests, 104, 105, 146
B.L.M.C. vehicles, body structure and computer analysis, 59, 144
BODY COMPONENTS, 50
Body motion and interior noise, relationship, 118, 146, 152
BODY STRUCTURES, EFFECTS OF PRESENT AND POSSIBLE FUTURE SAFETY LEGISLATION, AND THE MATHEMATICAL SIMULATION OF THE BARRIER IMPACT, 19
Bond *Bug*, plastics, 42
British Standard 857, windscreen glass, 44
Bumpers; 2
skinned micro-cellular urethane foam, 148, 153
Buses, structural analysis; bending, 75, 81, 82
Continental design, 71
integral, 71
load carried by bodywork, 71, 81, 82
loads in side-frame due to bending, 81, 82
matching body and chassis frame, 71
models, plastics, 76, 77, 79, 80
stress analysis, 71, 81, 82
torsional stiffness, 78, 82
underfloor grill, 71
underfloor structures, 74
window pillars, bending moments, 74, 75, 81

Cable, servo-controlled, impact test, 7
CADANCE design technique, 4, 148, 153
CHALLENGE OF AUTOMOBILE DESIGN FOR THE 1970s, 1
Chassis frame; and body matching, buses, 71
collision energy absorption, 90
corrosion, 149, 151
crushing machine, collision study, 91
function, 88, 149, 150
laboratory durability testing, 92
'perimeter' design, 90, 148, 150
road force simulation in tests, 91, 92, 94
soft body attachment, 89
CHASSIS FRAMES, 88
Chevrolet cars; bumpers, 153
plastics, 36, 41
Citroën cars, plastics, 38
Coaches, *see* Buses
Colours, interior, 145, 153
Comfort; location of controls, 99, 145, 151
seating, 86, 145, 153
windscreens, 47
Commercial vehicles, plastics, 35
Components, typical, 51
Computer-aided design; 4, 148, 153
advantages, 141
automatic drafting machine, 136
CADANCE technique, 4, 148, 153
clay model–integrated tape preparation system, 135
dynamic sketch pad and light pen, 4, 136, 139, 140
tool manufacture, 137
translation of clay model into body draft, 134, 146, 147, 148
Computer-aided structural analysis; 55, 143, 144
calculation times and programme make-up, 64, 69
comparison with test results, 67
cost, 58, 143, 144, 146, 147, 148
idealization, 59, 144
rear floor, 64
side frame, 66
Computer simulation, impact test, 4, 20, 143
Continental bus design, 71
Controls; driver's reach, 96, 145, 151
emergency, 145, 151
fitting trials, 98
layout criteria, 99, 145, 151
legibility and location, 3
separation, 98
standardization, 100, 146, 153
symbol system for identification, 86
visibility, 99, 145, 151
Corrosion; 'B' post, 146
chassis frames, 149, 151
Cost; computer-aided structural analysis, 58, 143, 144, 146, 147, 148
manikin, 145
single/multi-point vibration measurement, 146, 152
tooling, plastics, 42
Crushing machine, study of collision reaction of car frames, 91
Cybernetics, 85, 86
Cycolac research vehicles, plastics, 38, 42
Cylinders, impact tests for study of vehicle-impact behaviour, 27, 31

Damping, noise reduction, 130, 146, 149, 152
Deceleration curves, windscreens, 47
Deceleration measurement, impact tests, 7, 10, 11, 14, 21, 26, 142, 150
Deceleration severity, Injury Index, 22
Design, *see* Computer-aided design
Doors; drop tests, 106
hinges and attachments, 15, 16, 50
Doors, latches; American safety regulations, 50
and handles, development, 50
dovetail type, 52
production economics, 52
Doors; side, strength, 14, 16
window regulators, 50
Drafting machine, automatic, 136
Drawing by tape, 3
Drive shaft centre bearing, vibration, 112
Drivers; forces exerted by, 99, 145, 151
reaching controls, 96, 145, 151
reaction time, 98, 145, 151

'Dynamic sketch pad', 4
Dynapak principle applied to linear accelerator, 29

Economics; door latch production, 52
of use of plastics, 42, 143, 151
Engine cross member, vibration, 111
Ergonomic aspect of interior fittings, 96, 145, 151
Ergonomics and driver's environment, 85, 86, 145, 153
Ergosphere, reach distances, 97, 145, 151
Europe, Economic Commission for, safety regulations, 12, 19
EXPERIMENTAL INVESTIGATION OF BODY STRUCTURAL VIBRATION, 108

Finite-element analysis, 75, 143
Floor; rear, computer analysis, 64
separation from under-floor structure, 146, 152
Forces exerted by drivers, 99, 145, 151
Ford cars; *Cortina*, seating, 145, 153
side-door strength, 14
Ford Motor Company impact tests, 6
Front-engine vehicles, safety, 143
Fuel tanks; frontal impact tests, 9
safety regulations, 9, 12, 19

Gadd Severity Index, 23
General Motors; design aids, 3
design Policy Groups and Committees, 3
interior and exterior design, 1
Glass, windscreen; laminated, 45, 143
specification BS 857, 44

Head/seat restraint design, 14
Headlights, 2
Heater units, 2
Hillman cars; *Avenger* sill, bending tests, 146
plastics, 40
static tests, 102
Hinges, 15, 16, 50
Honda cars, plastics, 40, 143, 152
HUMAN FACTORS INFLUENCING CONTROL POSITIONS, 96

IMPACT BEHAVIOUR, STUDY OF THROUGH THE USE OF GEOMETRICALLY SIMILAR MODELS, 26
Impact tests; accelerometer traces, 7, 142, 150
anthropometric dummy 'Sierra Stan', 8
chassis frame reaction, 90
computer simulation, 4, 20, 143
cylinders, behaviour during crushing, 27, 31
deceleration measurement, 7, 10, 11, 14, 21, 142, 143, 150
effect of seat belts, 10, 12, 17, 22
forces, 24, 143
frequency response of transducer signals, 6, 142, 150
Impact tests, frontal; American standards, 9
windscreen retention, 9, 19, 20, 142, 150
Impact tests; Gadd index, 23
Injury Index, 22
Instron machine, 31
instrumentation, 6, 142, 150
intersection collision, 142, 150
linear accelerator, 29
occupant simulation, 8
potential future regulations, 20
Impact tests, rear; computer simulation, 24, 143
fuel-tank integrity, 12, 19, 20
seat/head restraint development, 14
servo-controlled cable, 7
Severity Index, 7, 10, 23
standard methods, 12
Impacts tests, side; computer simulation, 25, 143
door, strength, 14
effect of stiffening, 17, 142, 150
S.A.E. mobile barrier, 14
standard methods, 14
Impact tests; steering column horizontal penetration, 9, 19, 20
Vauxhall *Viva* station wagon, 21
windscreens, 47
with geometrically similar models, 26
Injury Index, 22

Instron machine, cylinder crushing tests, 31
Instrument panel; assembly and installation, 2, 86
design for safety, 3
Instruments, legibility and location, 3
Instrumentation; impact tests, 6
vibration measurement, 114
Insulation, noise, 130, 146, 152
Interior design; colours, 145, 153
ergonomic aspects, 96, 145, 151
future developments, 86
General Motors, 3
safety, 84, 145, 146, 153
See also Noise, interior

Jacking point, vibration, 112, 146
Joints, tests, 107

Layout, interior, ergonomic aspects, 99, 145, 151
Light pad and light pen, 4, 136, 139, 140
Lights, park/signal, 2
Linear accelerator, impact tests, 29
Load, bodywork of bus, 71, 81
Loading, bending tests, 105, 146
Lotus car, fibreglass, 35

MAMA apparatus, 124
Manikins; cost, 145
design, 8, 96, 145, 151
Marcos cars, fibreglass, 35
Market research, 148, 153
Market saturation for dictating style, 154
Materials; acoustical, 130, 133, 146, 152
steel, future shortage of raw materials, 34
See also Plastics
Mobility and interior noise, relationship, 118
Mobility testing, vibration, 108, 146, 152
Model tests, vehicle-impact behaviour, 26
Models, clay; 3, 4, 135
computer method of translation into body draft, 134
co-ordinate information, measurement and recording, 135, 136
integrated tape preparation system, 135
Models, plastics, buses, structural analysis, 76, 77, 79, 80
MODERN METHODS OF TRANSLATING A STYLING MODEL TO A BODY DRAFT AND TOOLING, 134
Mounting vibration, 112, 120

Noise, interior; absorption, 133, 146, 152
'acoustical holes', 132
acoustical materials, 130, 133, 146, 152
assessment, 127, 146, 152
damping, 130, 146, 152
reduction, investigation techniques, 128, 129, 146, 152
relationship with body motion, 118
relationship with speed, 127
resonance phenomena, 128
sound insulation, 130, 146, 149, 152
transmission modes, 130, 146, 152
vibration insulation, 130, 146, 152
Noise, tyre, 149

Occupant simulation, impact tests, 8

Painting, plastics, 143, 152
Panel, vibration, 111, 112, 128, 130
Perimeter frame, 90, 148, 150
Petrol filler pipe location, 2
Photography, impact tests, 6, 8, 15, 16, 26, 31
Plastics; A.B.S. materials, 40
advantages, 34
blow moulding, 36
compression moulding, 36
economics, 42, 143, 151
fibre-reinforced polyesters, 41
growth-rate, past and projected, 34
I.C.I. patented process, 41
injection casting, 38
injection moulding, 35
material wastage in moulding, 34

models, buses, structural analysis, 76, 77, 79, 80
nylons, 41
painting, 143, 152
polypropylenes (thermoplastics), 39
polyurethanes, 41
production rate, 42, 143, 152
radiator grille, 40, 143
rotational casting, 37
tooling costs, 42
vacuum forming, 38
Pontiac cars, bumpers, 153
Power operated window regulator, 51
Production processes, plastics, 35, 143, 152

Radiator grille, plastics, 40, 143
Reach of drivers, 96, 145, 151
REALISM OF VEHICLE IMPACT TESTING, 6
Reliant cars, fibreglass, 35
Road forces, simulating machines, 91, 92, 94
Road intersection collision test, 142, 150
Road resistance, windscreens, 47
Roll-over test, 20
Roof, effect on stiffness, 144
Rootes test rig, 103
RTAD apparatus, vibration study, 128

S.A.E. mobile barrier, impact tests, 14
Safety; 'air bag', 3, 84
collapsible steering wheel, 85
collision reaction of chassis frames, 90
design of instrument panel, 3
driver's environment, 85, 86, 145, 153
ergonomics, 85, 145
front-engine vehicles, 143
fuel tanks, Economic Commission for Europe, regulations, 12, 19
interior design for, 1, 84, 145, 146, 153
its challenge to designers, 84
Safety regulations; American, 9, 12, 19, 50
potential future, 20
Sweden, 19
Safety seat, 85
Safety; side stiffening, 17, 142, 150
windscreen glass, 45, 47
Seat belts; anchorage tests, 106
effect in collisions, 10, 12, 22
Seat/head restraint design, 14
Seating; comfort, 86, 153
Ford *Cortina*, 145, 153
noise level at, 117, 119, 120
vibration, 115, 120
Servicing, interior of car, 2
Severity Index, impacts, 7, 10, 23
Shakers, hydraulic, chassis frame tests, 149, 151
Shelby *Mustang* safety seat, 85
Side, exterior, protection, 2
Side frame; buses, loads due to bending, 81, 82
computer analysis, 66
Side stiffening, 17, 142, 144, 150
'Sierra Stan' anthropometric dummy, 8
Sill; *Avenger*, 146
design, 15, 16, 74, 143, 144
Simca body shell, bending tests, 104
SMALL COMPUTER PROCEDURES AS TOOLS FOR STRUCTURAL DESIGNERS, 62
SOUND DAMPING CONTROL OF AUTOMOBILE STRUCTURES, 126
Speed; impact, relationship with vehicle deformation, 10
relationship with interior noise level, 127
Springs, vibration, 116, 117, 120
Stabilizer bracket, vibration, 109
Standardization of controls, 100, 146, 153
Static tests, 102
Steel production, future shortage of raw materials, 34
Steering column; collapsible, 85
frontal impact tests, 9
safety standards, 9, 19
Stiffening, side body, 17, 142, 144, 150
Stiffness, effect of roof, 144
Stress analysis, bus structures, 71, 81, 82
Structural analysis, computer-aided; comparison with test results, 67
cost, 58, 143, 144
idealization, 59
programme, examples, and calculation time, 63, 64, 69
rear floor, and side frame, 64, 66
STRUCTURAL DESIGN OF BUS BODIES, 71
Struts, failure analysis, 28
Styling; dictating by market saturation, 154
windscreens, 48
Suspension; development, 89
vibration, 112, 120
Sweden, safety regulations, 19
Switches, time needed to operate, 99
Symbol system, identification of controls, 86

Test rig, A.S.A.E., for verification of structural analysis by computer, 68
Tests; bending, 104, 105, 146
door drop, 106
for verification of computed structural analysis, 67
impact, *see* Impact tests
joints, 107
laboratory, chassis frame durability, 92
roll-over, 20
seat belt anchorage, 106
static, Rootes rig, 102
'Trauma-saf' device, 3
Tool manufacture, computer-aided design, 137
Tooling costs, plastics, 42
Torsion tests, 105
Torsional stiffness, buses, analysis, 78, 82
TOWARDS THE ALL-PLASTICS MOTOR CAR, 34
Transducers, impact tests, 6, 142, 150
'Trauma-saf', safety tests, 3
Tubes, rectangular, failure analysis, 28
Tyres, noise tests, 149

Underfloor; grill, buses, 71
separation from floor, 146, 152
structures, buses, 74

Vauxhall *Viva* station wagon impact tests, 21
Vibration; insulation, 130, 146, 152
interior, noise and mobility relationship, 118
separation of floor from under-floor, 146, 152
Vibration measurement; cost, single/multi-point systems, 146, 152
instrumentation, 114
Kennedy and Pancu method, 108, 146, 152
MAMA apparatus, 124
single-point excitation technique, 108
the case for the multi-point system, 108, 124
Vibration study, RTAD apparatus, 128
Visibility, of controls, 99, 145, 151
Vision; forward, 2
rear, 3
windscreen glass, 44, 48

Wheel spindles, recording movement, 93
Window regulators; 50, 143
power operated, 51
Windows; effect of glass on noise and vibration, 120
pillars, bending moments, 74, 75, 81
WINDSCREENS OF THE FUTURE, 44
Windscreens; comfort, 47
deceleration curves, 47
design for safety, 3, 45, 47
effect on noise and vibration levels, 120
environmental conditions, 48
future requirements, 44
glass, specification BS 857, 44
laminated glass, 45, 143
retention in impact, 9, 19, 142, 150
road resistance, 47
styling, 48
vehicle performance, 47
vision, 44, 48

MADE AND PRINTED IN GREAT BRITAIN BY WILLIAM CLOWES & SONS, LIMITED, LONDON, BECCLES AND COLCHESTER